Synthesis Lectures on Operations Research and Applications

This series focuses on the use of advanced analytics in both industry and scientific research to advance the quality of decisions and processes. Written by international experts, modern applications and methodologies are utilized to help researchers and students alike to improve their use of analytics. Classical and cutting-edge topics are presented and explored with a focus on utilization and application across a range in practical situations.

Yusuf Ihsan Tokel · Jun Zhuang

Optimizing Security Patrolling Strategies

A Cross-Domain Review of Mathematical Models and Applications

Springer

Yusuf Ihsan Tokel
Department of Industrial and Systems Engineering
Buffalo State University
Amherst, NY, USA

Jun Zhuang
Department of Industrial and Systems Engineering
Buffalo State University
Buffalo, NY, USA

ISSN 2770-6303 ISSN 2770-6311 (electronic)
Synthesis Lectures on Operations Research and Applications
ISBN 978-3-032-02616-3 ISBN 978-3-032-02617-0 (eBook)
https://doi.org/10.1007/978-3-032-02617-0

This Springer imprint is published by the registered company Springer Nature Switzerland AG
The registered company address is: Gewerbestrasse 11, 6330 Cham, Switzerland

Contents

Acronyms

AIS	Artificial Immune System Algorithm
ALNS	Adaptive Large Neighborhood Search
AMCL	Adaptive Monte Carlo Algorithm
APCP	Area-Partition-Involved Collaborative Patrolling
ASAP	Agent Security via Approximate Policies
CBLS	Concurrent Bayesian Learning Strategy
CDQN	Censoring Deep Q-Network
COBRA	Combined Observability and Rationality Assumption
CP	Constraint Programming
DDQN	Double Deep Q-Network
DFS	Depth-First Search
DOBBS	Decomposed Optimal Bayesian Stackelberg Solver
DP	Dynamic Programming
DQN	Deep Q-Network
DRL	Deep Reinforcement Learning
DT-DRL	Diverse Trajectory-Driven Deep Reinforcement Learning
EBSH	ϵ-Box Splitting Heuristic
ECH	ϵ-Constraint Heuristic
ERASER-C	Efficient Randomized Allocation of Security Resources with Constraints
FB-SAMCP	Factored Belief Based Sequential Allocated Monte Carlo Planning
LP	Linear Programming
MC	Markov Chain
MDLS	Multi-directional Local Search
MDP	Markov Decision Process
MIDAS	Maximin Defense Against Subjective Utility Quantal Response
MILP	Mixed Integer Linear Programming
NISFA	Non-dominated Immune endocrine Short Feedback Algorithm
ORIGAMI	Optimizing Resources in Games using Maximal Indifference

POMDP	Partially Observable Markov Decision Processes
PSDA	Probabilistic Solution Discovery Algorithm
TIAOR	Time-Priority Immune Clonal Selection Algorithm with Optimization-Revision
U-PBCPA	UAV-Police Vehicle Cooperative Patrol Algorithm
UANOA	USVs Autonomous Navigation and Obstacle Avoidance
UAV	Unmanned Air Vehicle
USV	Unmanned Surface Vehicle
WTSC	Weighted Targets Sweep Coverage

1 Introduction

1.1 Context

Patrolling is one of the most effective and widely adopted crime prevention and deterrence strategies worldwide. It is integral to security agencies such as police and military forces across various domains, including land, air, and maritime areas. Effective patrolling requires the coordination of manpower, technological resources, and policies to address evolving security challenges. Research on patrolling spans multiple fields, particularly in the domain of Operations Research. This field has extensively studied optimization techniques to enhance the allocation of patrol resources, reduce response times, and maximize deterrence effects.

Recently, there has been an increasing trend of research addressing patrolling problems in different domains. Robotic patrolling in particular has received considerable attention, with numerous studies exploring its challenges and applications. Robin and Lacroix [33] introduced a taxonomy-oriented framework to classify multi-robot missions and associated problems, providing a systematic approach to relate findings from diverse research communities under a unified taxonomy. Huang et al. [19] conducted an in-depth analysis of the literature on multi-patrol strategies, categorizing studies into two primary patrolling types: regular and adversarial. The focus was exclusively on multi-robot scenarios, providing a comparative examination of these two approaches. The study highlighted the strategic distinctions, where regular patrolling emphasizes coverage and routine monitoring, while adversarial patrolling prioritizes countering threats or intrusions in dynamic and potentially hostile environments. Recently, [3] discussed advancements and open questions in the robotic patrolling domain, highlighting key research directions. These include extracting instances from real-world cases for simulation, emphasizing tailored modeling to improve solution quality over generality, exploring long-term reliability to enhance robots' operational longevity, increasing focus on adversarial settings, and integrating various components to improve overall solution effectiveness.

Y. I. Tokel and J. Zhuang, *Optimizing Security Patrolling Strategies*, Synthesis Lectures on Operations Research and Applications, https://doi.org/10.1007/978-3-032-02617-0_1

On the police patrolling scope of the problem, two review papers are worth mentioning, considering their relevance to this review. Dewinter et al. [15] examined the police patrol routing problem (PPRP) and the dynamic vehicle routing problem (DVRP), analyzing 30 studies to compare their solution methodologies, algorithmic strengths and weaknesses, and modeling approaches. The study identified hybrid genetic algorithms, routing policies, and local search methods as the most effective techniques for addressing these challenges. Samanta et al. [35] conducted a comprehensive literature review on police patrolling from an operations research perspective. They emphasized the practicality of the problems and prioritized models that offer implementable solutions. The study categorized problems based on their modeling approaches, a framework that this review also adopted and adapted to address emerging demands identified during the review process. However, to the best of our knowledge, no study in the literature offers a thorough and systematic analysis of patrolling problems that covers robotic, autonomous, human, and hybrid approaches, while also providing a detailed, statistically grounded summary of research directions. To bridge these gaps, this study offers a comprehensive overview of the different patrolling approaches and highlights their significance, especially within security-related applications.

1.2 Reactive Versus Proactive Policing

We begin with a broader classification that is commonly found in the literature. Although this classification extends beyond patrolling operations, it is frequently used to distinguish between different policing frameworks. In practice, proactive and reactive policing approaches are closely linked to the various types of patrolling strategies commonly employed. The term "proactive" in terms of security applications first coined by [6], in the context of centralized and decentralized organizational structures, indicating that in order to perform proactive policing, more professional organization would be required to give the necessary autonomy to security officers, with avoiding the risk of corruption originating from lack of centralized organizational structure. Although the implementation of proactivity was observed through historical reports (in the form of developing networks of criminals, offering immunity from arrest for information during the early twentieth century, as stated in [18]), it was not institutionalized as we understand proactivity today. The state-of-the-art policing method was referred to in [14] as a "standard" model and described as "all-round reliance on random patrol, rapid response to calls for service, follow-up investigations by detectives and unfocused enforcement efforts." As the definition also hints, the focus on standard approach is being reactive to the reported incidents: the aim is to reach out the incident location as early as possible. The random patrolling, which aimed to increase the omni-presence of the security forces, was also utilized to shorten the time before arriving to an emergency call, or reacting to a reported incident. 1960s could be seen as the turning point to question standard policing methods in the US as the opposition to Vietnam War grown, unrest in minority groups towards racial injustice escalated among younger Americans and the absolute need for serious improvements of the standard model started to be discussed seriously in 1970s [29].

The first, maybe the most influential, scientific study to measure the effectiveness of standard policing methods was conducted by [21], a lengthy experiment which lasted from October 1, 1972 to September 30, 1973. They divided the city into three experimental areas: in one area, they only send police troops only upon receiving a call, in the second area, they doubled, sometimes tripled the police presence, and they treated the third area as a control group and performed patrolling duties as usual. The conclusion was surprising: they did not report any statistically significant difference in terms of the level of crime, citizens' perception of police services, citizens' fear of crime, response time, or citizens' satisfaction with response time. But later clearly concluding that this conclusion must not be evaluated as the police presence does not have any effect on crimes in general, but traditional methods do not work as it is assumed to be. However, this study is often falsely cited to strengthen the idea of "police presence does not deter" as [39] also pointed out. Sherman and Weisburd [39] also showed in his study that this experiment cannot be generalized to have such strong claims because of the problems in its design, experimental conditions, inherent bias, to name a few. They claimed that in order to overcome these experimental design problems, the studies must consider very small clusters of high-crime addresses, instead of large patrol beats or neighborhoods.

During the 1970s, some studies objected the outcomes of [21] and reported more positive results when employing standard, more reactive patrolling methods. Chaiken [9] mentioned that there are multiple types of deterrence and some measures are extremely difficult to measure because of the fact that trying to measure something that does not happen. And in [21] they only measure one aspect of deterrence, and since they were not able to measure other deterrence effects, we cannot claim that traditional policing practices, or more police presence in the neighborhoods can have no effect, which also agrees with [39] from a methodological perspective. Along similar lines, [36] also raised similar concerns about the credibility of the [21] study and by designing another experiment in Nashville, focusing more on the difference on day/night shifts, they claimed that in order to maintain reduced level of crime, increasing patrol levels, by increasing both staff and other relevant resources, might be necessary.

More recent studies also suggested to not simply abandon standard police methods because assuming proactive methods are more superior. In order to show the importance of reactive patrolling methods, [41] analyzed the level of crimes when police strikes were conducted. To seek an answer to the hypothesis "the more police a city employs, the less crime it will have" [41] divided this question into two parts. First, to measure the absolute effect of police on preventing crimes, they reviewed prior studies in which researchers studied the crime levels before, during and after strikes. As expected, the authors reported that crimes increased dramatically during the strikes in 5 of these 6 studies. The one study concluded not a significant increase in crimes during a strike actually was poor in design because 89% of the days in strike period actually covered non-strike days. Second, to measure the marginal effect of additional police officers, they reviewed prior review studies and concluded that the results are not as powerful as the first one: some studies found little evidence that more police reduce crime, whereas other studies did not report any statistically significant difference on reducing crimes between having more police officers or not. Nagin and Weisburd [28] also reviewed different experimental cases to analyze the effect of increasing police presence on crime rates and in one of the studies they reviewed [22] increasing police presence did cause a significant drop in crime levels.

Although standard methods remain effective and are still widely used, proactive policing techniques have emerged to enhance the ability to address crime and disorder more broadly. As defined by [29], proactive policing includes "all policing strategies that have as one of their goals the prevention or reduction of crime and disorder and that are not *reactive* in terms of focusing primarily on uncovering ongoing crime or on investigating or responding to crimes once they have occurred." They categorized the proactive policing into four main categories, as place based approach, problem-solving approach, person-focused approach and community-based approach. Among all these categories, we want to pay a special attention to place based approach because, as [48] pointed out, place-based approach is more efficient than person-based approach especially with regards to the resource allocation point of view: Since there is a vast majority of evidence to support the idea that a great majority of the crimes actually occur within a small portion of places [39, 42], allocating limited security resources to these high-risk areas would maximize the efficiency. Therefore, among others, hot-spot policing/patrolling strategies are the most adopted strategies within practitioners. It's important to note that proactive policing strategies often overlap in practice. For example, during a shift, an officer working a hot spot might also engage in community interactions–informing residents about ongoing concerns while gathering information about local incidents. This blend of place-based and community-based techniques reflects the integrated approach highlighted in [46].

1.3 Types of Patrolling

Literature shows distinct comparisons based on different objectives and methodologies regarding patrolling operations. In this study, we will be comparing two such classifications to make patrolling more understandable within different contexts.

1.3.1 Random Versus Directed

Although the idea of concentrating limited security resources to the places that are more crime-prune than other places sounds old and well established, as we introduced in the previous section, the crime prevention strategies did not methodologically focus on concentrated crime locations until 1970s [17]. Instead, more reactive approaches were widely adopted, such as random patrolling.

With random patrolling strategy, an officer aims to be unpredictable in terms of patrolling patterns, aiming to generate an overall omnipresence to deter criminals to attempt any criminal activity [40]. Within this context, randomization refers to a pure randomness without following any hot spot, historical patterns; but relying on patroller's own experience. Whereas directed patrolling refers to a more structured strategy where the patroller increase their presence in hotspots during hot times based on historical crime patterns [16]. The theory behind the strategy often points out how the crimes within a neighborhood is concentrated in particular areas and increasing security officer presence at these locations will reduce the crime substantially compared to pure random patrols. One of the main concerns of directed patrolling is that over time, once the security presence will be predictable at certain locations (hot-spots), would that cause the crime displacement to seemingly low crime locations. In a recent study by [7] concluded that patrols concentrated on hot-spots generated statistically significant small reductions in overall crime and does not produce crime displacement effects (as also concluded by [14]), rather it causes the crime to diffuse nearby immediate surrounding target locations.

While directed patrolling in general, hot-spot patrolling in specific has generally shown greater effectiveness in reducing crime [7, 39], it has also faced significant criticism. In particular, it is often viewed as a contributing factor to police-initiated violations, especially against non-white communities, as noted by [29]. However, evaluating this dimension of policing strategies lies beyond the scope of this research. For a more in-depth analysis, we refer interested readers to [30].

1.3.2 Regular Versus Adversarial

A different type of categorization can also be made purely on patrolling strategies than the previous two subsections. Adversarial patrolling term first coined by [1] in a multi-robot patrolling setting. The comparison between a regular (frequency-based) and adversarial (probability-based) patrolling can be defined as follows: in a regular patrolling strategy, the objective is to visit an area of interest as frequently as possible, in order to maximize the visibility and minimize the uncovered time of space in general. Adversarial patrolling, in contrast, aims to maximize the probability of detection, or capture [8, 13, 26, 31, 45]; or minimize the non-detection probability [25, 34] depending on the context. But the objective is very different, and the frequency of visits becomes less important in adversarial scenarios.

Although this categorization is primarily used in robotic patrolling scenarios, we argue that the underlying distinction is equally applicable to non-robotic patrolling contexts. When the objective relates to the frequency of visits–such as maximizing coverage [10, 11, 37], minimizing idle time on the graph or at specific nodes [32, 49], or increasing police visibility rates [47]–we classify these models as regular patrolling, regardless of whether they involve robots, unmanned vehicles, or human agents. Drawing inspiration from [19], who categorized models based on patrolling type preferences in a recent survey of multi-robot patrolling literature, we extend this framework to encompass all patrolling models analyzed in the following sections.

1.4 Modeling Themes

In our literature review, we classified the models employed in the examined studies into three main categories: deterministic, stochastic, and game-theoretic. In this section, we introduce each category within the scope of this study and highlight their defining characteristics. Despite their differences, these modeling approaches share several foundational elements:

Decision variables: All models require decision variables to represent the choices or actions taken under certain conditions and constraints. These variables can be continuous (e.g., real numbers), discrete (e.g., integers), or binary. The type of decision variables significantly influences both the formulation and the solution method. For instance, if all variables are continuous and the relationships are linear, the problem may be solved using Linear Programming techniques such as the simplex method or column generation. However, if the model includes both integer or binary and linear variables, it becomes a Mixed-Integer Linear Program, requiring different, more advanced tools like branch-and-bound, branch-and-price, or Benders decomposition to solve.

Objective function: Every model seeks to optimize an objective function, which mathematically represents the goal of the decision-making process. This objective can be to minimize (e.g., cost, distance, risk) or maximize (e.g., profit, coverage, visit frequency) a certain metric. In patrolling-related applications, for example, objectives might include maximizing the frequency of visits to certain areas or minimizing the total distance traveled. In some cases, models involve multiple conflicting objectives, which require the use of multi-objective optimization techniques, are discussed in subsequent sections.

Constraints: Constraints express the limitations of available resources or operational rules. These are essential in modeling because they define the feasible region for the decision variables. Common constraints include budgets, capacities, time windows, and coverage requirements. The number and complexity of constraints often correlate with the computational difficulty of a problem: in general, more constraints lead to a more challenging and time-consuming solution process. However, it is important to note that this is not always the case. In some instances, adding well-defined constraints can actually reduce computational time–particularly when they help the solver generate stronger cuts, tighten relaxations, or better bound the feasible region. Although it is highly case-dependent, such constraints can guide the solver more effectively through the solution space and improve convergence.

1.4.1 Deterministic Models

Deterministic models are defined by certainty: every input parameter is known exactly, and there is no randomness or uncertainty in the system. As a result, the solution computed –whether via linear programming, nonlinear programming, integer programming, or mixed-integer programming—is guaranteed to be a global optimum within the specified constraints and objective for convex problems. For non-convex optimization problems though, the global optimum is not guaranteed and there are several ways to obtain it using various techniques, such as signomial programming or general convexification strategies, which are reviewed by [27] in detail.

In practice, many real-world problems are so large or complex that finding an exact optimum within a reasonable time is intractable. As a result, practitioners often turn to heuristic or metaheuristic algorithms to generate high-quality, "good-enough" solutions quickly. The performance of these approaches is typically evaluated using two key metrics: Solution time is the required time for the algorithm to produce a solution. Shorter solution times indicate greater computational efficiency. Optimality gap is the relative difference between the heuristic solution's objective value and the true (or best-known) optimal value, usually expressed as a percentage, in which a smaller gap means the heuristic solution is closer to optimal, which is critical when the true optimum is known or tightly bounded. Practitioners often balance these two competing objectives–solution quality versus computation time–when selecting an approach for large-scale problems. For instance, if a metaheuristic yields a 2% optimality gap in 10 min, whereas obtaining the exact optimum for the same instance requires several hours, a decision-maker may favor the metaheuristic despite its slight suboptimality.

In our literature review, we labeled a study as employing deterministic optimization if it met all of the following criteria:

- The model does not include any uncertain or probabilistic inputs, i.e. all the parameters and constraints are fixed and known
- The objective function is a direct mathematical expression (e.g., linear, nonlinear, integer) rather than an expectation
- Does not include any scenario or chance-based constraints, all the constraints are fixed and hold deterministically
- Re-solving the same instance must yield to the same solution.
- The core formulation employs one of the following deterministic solution frameworks–either an exact programming class (LP, NLP, BP, MIP, MILP, MINLP, DP or CP) or a deterministic heuristic/metaheuristic (e.g., genetic algorithms, simulated annealing, tabu search).
- All decisions are made by either a single centralized optimizer or a group of cooperating agents using non-strategic, non-adversarial coordination (e.g., rule-based protocols or auction mechanisms)

1.4.2 Stochastic Models

Before defining stochastic models, we first need to identify the difference between two interrelated terms non-deterministic and stochastic. A stochastic model is a non-deterministic model but not all non-deterministic models are stochastic. So, with deciding with the name stochastic models, we are excluding some theoretical models that might not be neither deterministic nor stochastic, but non-deterministic. However, from the papers we analyzed, all papers adopted non-deterministic approaches were actually employing stochastic optimization, that is randomness is explicitly introduced and modeled. Therefore, instead of using the broader term non-deterministic, we chose to use stochastic as our categorization label, as it more precisely reflects the modeling approaches observed in our review and aligns with the terminology more commonly adopted in the operations research and management science (OR/MS) literature (see [5, 38]).

In its simplest form, stochastic optimization refers to procedures for maximizing or minimizing an objective function when some inputs or system behaviors are uncertain and modeled as random variables [44]. As discussed, stochastic optimization sets the stage for models where uncertainty is explicit. A foundational class under this umbrella is Markov processes, which assume memoryless state transitions and they extensively used in the patrolling literature. Therefore, in the following sections, we briefly introduce decision-making under uncertainty through Markov Decision Processes (MDPs), and further into Partially Observable MDPs (POMDPs) when the agent cannot directly observe the state.

Markov Chains: A Markov chain is a tuple $< S, P >$ where S is the set of possible states and P is the state-transition probability matrix. Its defining feature–the Markov property–asserts that the probability of moving to a new state depends solely on the current state, not on the sequence of past states. This memoryless property makes Markov chains both powerful and elegant for modeling random dynamics and it underpins richer frameworks like MDPs, which add decision and reward structures, and POMDPs, which further account for partial state observability under uncertainty.

Markov Decision Processes: An MDP extends the Markov chain framework by introducing actions and rewards. An MDP is given by the tuple $< S, A, P, R, \gamma >$ where S denotes set of states, A the set of available actions, $P(s' \mid s, a)$ the probability of transitioning to state s' after taking action a in state s, $R(s, a)$ the expected immediate reward for that action-state pair, and γ the discount factor balances the immediate versus future rewards. An MDP is called discrete if the state and action spaces are discrete.

A policy in this context is any rule that assigns to each state a (possibly randomized) choice of action, and the objective is to find the policy that maximizes the expected sum of discounted rewards over time. Classic solution methods built on Bellman's principle of optimality [4] include value iteration and policy iteration [20]. In a patrolling scenario, for example, locations can be modeled as states, movement decisions as actions, and transition probabilities as the likelihood of successfully traversing between locations; the reward function might then reflect metrics such as coverage frequency or minimized idle time, and solving the MDP yields an optimal routing strategy under uncertainty.

Partially-observable Markov Decision Processes: While an MDP assumes that the decision-maker has perfect knowledge of the current state, many real-world problems–such as autonomous vehicle patrolling–feature imperfect or noisy observations. A POMDP addresses this by extending the MDP tuple $< S, A, P, R, \gamma >$ to $< S, A, P, R, \gamma, Z, O, b_0 >$ where Z is a set of observations, $O(z|s', a)$ gives the probability of receiving observation z after taking action a and landing in state s', and b_0 is the decision-maker's initial belief, or probability distribution over states. After each action and observation, the agent updates its belief about the true state and selects the next action according to a belief-based policy that seeks to maximize expected cumulative reward.

Smith and Simmons [43] mentions exact value-iteration algorithms for POMDPs, but these methods suffer from an exponential explosion in representation size as the number of states grows. In practice, most large-scale POMDPs are solved approximately using point-based or other heuristic methods that focus computation on the most relevant regions of the belief space and aim to minimize expected regret starting from b_0. Recently, [24] surveyed the POMDP literature in great detail and we refer interested readers to this comprehensive study.

In our literature review, we labeled a study as employing stochastic optimization if it met all of the following criteria:

- The model includes uncertain or probabilistic inputs, such as random variables, stochastic parameters, or data drawn from known or estimated probability distributions.
- The objective function explicitly accounts for uncertainty, typically through the use of expectations, risk measures (e.g., variance, CVaR), or probabilistic cost functions.
- The formulation includes scenario-based representations, stochastic constraints, or chance constraints that only need to hold with a certain probability level.
- Re-solving the same instance with different random seeds, samples, or scenarios may yield different solutions due to the inherent randomness modeled within the problem.
- The core formulation belongs to a recognized stochastic optimization framework–such as two-stage or multi-stage stochastic programming, chance-constrained programming, sample average approximation (SAA), or sequential decision-making models like Markov Decision Processes (MDPs) and Partially Observable Markov Decision Processes (POMDPs)–and is solved using either exact methods or stochastic/metaheuristic approaches tailored to uncertainty.
- All decisions are made by either a single centralized optimizer or a group of cooperating agents using non-strategic, non-adversarial coordination (e.g., rule-based protocols or auction mechanisms).

1.4.3 Game-Theoretic Models

The final category in our model-based classification is Game Theory. Throughout the review, game-theoretical studies were by far the most adopted primary modeling theme. In this section, we aim to briefly introduce game-theoretic modeling practices and highlight their relevance in operational and security-oriented decision-making.

Game theory provides a mathematical framework for analyzing situations where multiple decision-makers (players) interact strategically. Unlike the deterministic and stochastic optimization models we defined earlier–where decisions are made by a single centralized optimizer or through non-strategic coordination–game-theoretic models involve decentralized decision-making among multiple self-interested agents whose actions jointly determine outcomes. Although these models often rely on deterministic or stochastic optimization tools to compute best responses, equilibria, or strategy profiles, what distinguishes them is their strategic structure: each agent optimizes their objective while anticipating the decisions of others. This interdependence leads to equilibrium-based formulations rather than centralized optimization.

A foundational distinction arises between normal-form games, where strategies are chosen simultaneously without knowledge of others' decisions, and extensive-form games, where players move sequentially according to a structured game tree. The most fundamental equilibrium concept across both representations is the Nash Equilibrium (NE)–a strategy profile where no player can benefit by unilaterally deviating from their strategy given the strategies of others. In simultaneous-move games with incomplete information, where players have private types or beliefs about others, the appropriate extension is the Bayesian Nash Equilibrium (BNE). Here, each player maximizes expected utility based on their beliefs about others' types, under a common prior.

In sequential games, different refinements of Nash equilibrium are applied depending on the information structure. In games with perfect and complete information, the Subgame Perfect Nash Equilibrium (SPNE) refines Nash equilibrium by eliminating non-credible threats, requiring that players' strategies constitute a Nash equilibrium in every subgame. When games are sequential and involve incomplete information, players must form and update beliefs about others' types or actions, leading to the use of Perfect Bayesian Equilibrium (PBE)–a refinement that combines sequential rationality and belief consistency.

There exist even more stringent refinements of PBE. For example, the Sequential Equilibrium (SEq) [23] requires that beliefs be derived as the limit of fully mixed (trembling-hand) strategies, thus narrowing the set of acceptable off-path beliefs. The Intuitive Criterion [12] further restricts equilibria in signaling games by ruling out off-path beliefs that assign positive probability to types for whom a deviation would be strictly worse, regardless of the receiver's response. The Divinity Criterion [2] strengthens this further by prioritizing beliefs in favor of sender types that benefit most from the deviation, providing a more refined belief update rule for off-path moves. Although these refinements vary in strictness, they all apply to extensive-form games with asymmetric information. For the purposes of this review, we group these under the broader umbrella of PBE-based models, as they share common strategic assumptions and modeling structures.

A special class of sequential games involves leader and follower dynamics, commonly modeled as Stackelberg games. In these, one player (the leader) commits to a strategy first, and the other (the follower) observes this and best-responds. The resulting solution is known as a Stackelberg Equilibrium (SE). If the follower's type is uncertain and drawn from a known distribution, the model extends to a Bayesian Stackelberg Equilibrium (BSE), where the leader commits to a strategy that maximizes expected utility over all possible follower types.

In our literature review, we labeled a study as employing game-theoretic modeling if it met all of the following criteria:

- The model includes multiple decision-makers (players), each with distinct objectives, acting in a strategic environment where outcomes depend on the joint actions of all players.
- The decisions are decentralized: each player optimizes their own utility (or minimizes their own cost), potentially subject to the anticipated strategies of others.
- The solution concept is rooted in equilibrium analysis, such as Nash equilibrium, Stackelberg equilibrium, subgame-perfect equilibrium, or perfect Bayesian equilibrium.
- The problem formulation explicitly accounts for strategic interdependence and models incentives, adversarial behaviors, or belief updates (in the case of incomplete information).
- The core structure aligns with standard game-theoretic models–e.g., normal-form, extensive-form, Stackelberg, or Bayesian games–and uses solution techniques (e.g., best-response dynamics, bi-level optimization, mathematical programming with equilibrium constraints) that are tailored to game analysis.

1.5 Organization

The rest of the paper is organized as follows: Sect. 2 summarizes the methodology adopted in this study; Sect. 3 presents all findings based on several categories within mathematical optimization domains; Sect. 4 analyzes the future research directions of all the papers based on predefined categories and the frequency of themes; and finally, Sect. 5 summarizes the findings and analysis, and provides concluding remarks.

References

1. Agmon, N., Kaminka, G. A., and Kraus, S. (2011). Multi-robot adversarial patrolling: Facing a full-knowledge opponent. *Journal of Artificial Intelligence Research*, 42:887–916.
2. Banks, J. S. and Sobel, J. (1987). Equilibrium selection in signaling games. *Econometrica*, 55(3):647–661.
3. Basilico, N. (2022). Recent trends in robotic patrolling. *Current Robotics Reports*, 3(2):65–76.
4. Bellman, R. (1966). Dynamic programming. *Science*, 153(3731):34–37.
5. Birge, J. R. and Louveaux, F. (2011). *Introduction to Stochastic Programming*. Springer Publishing Company, Incorporated.
6. Bordua, D. J. and Reiss, A. J. (1966). Command, control, and charisma: Reflections on police bureaucracy. *American Journal of Sociology*, 72(1):68–76.
7. Braga, A. A., Turchan, B., Papachristos, A. V., and Hureau, D. M. (2019). Hot spots policing of small geographic areas effects on crime. *Campbell Systematic Reviews*, 15(3):e1046.
8. Brown, G., Carlyle, M., Salmerón, J., and Wood, K. (2006). Defending critical infrastructure. *Interfaces*, 36(6):530–544.
9. Chaiken, J. M. (1976). What's known about deterrent effects of police activities. Report, RAND Corporation.
10. Chen, X. (2012). Fast patrol route planning in dynamic environments. *IEEE Transactions on Systems, Man, and Cybernetics Part A:Systems and Humans*, 42(4):894–904.
11. Cheng, L., Zhong, L., Tian, S., and Xing, J. (2019). Task assignment algorithm for road patrol by multiple uavs with multiple bases and rechargeable endurance. *IEEE Access*, 7:144381–144397.
12. Cho, I.-K. and Kreps, D. M. (1987). Signaling games and stable equilibria. *The Quarterly Journal of Economics*, 102(2):179–221.
13. Chung, H., Polak, E., Royset, J. O., and Sastry, S. (2011). On the optimal detection of an underwater intruder in a channel using unmanned underwater vehicles. *Naval Research Logistics*, 58(8):804–820.
14. Council, N. R. (2004). *Fairness and Effectiveness in Policing: The Evidence*. The National Academies Press, Washington, DC.
15. Dewinter, M., Vandeviver, C., Beken, T. V., and Witlox, F. (2020). Analysing the police patrol routing problem: A review. *ISPRS International Journal of Geo-Information*, 9(3):157–17.
16. Eck, J. E. and Maguire, E. R. (2000). *The Crime Drop in America*, chapter Have changes in policing reduced violent crime? Cambridge University Press.
17. Eck, J. E. and Maguire, E. R. (2018). *Unraveling the Crime-Place Connection, Volume 22: New Directions in Theory and Policy (1st ed.)*, chapter Beyond Putting "Cops on Dots": Applying Theory to Advance Police Responses to Crime Places. Routledge.
18. Haller, M. H. (1976). Historical roots of police behavior: Chicago, 1890-1925. *Law & Society Review*, 10(2):303–323.
19. Huang, L., Zhou, M., Hao, K., and Hou, E. (2019). A survey of multi-robot regular and adversarial patrolling. *IEEE/CAA Journal of Automatica Sinica*, 6(4):894.
20. Kalyanam, K., Chandler, P., Pachter, M., and Darbha, S. (2012). Optimization of perimeter patrol operations using unmanned aerial vehicles. *Journal of Guidance, Control, and Dynamics*, 35(2):434–41.
21. Kelling, G. L., Pate, T., Dieckman, D., and Brown, C. E. (1974). The kansas city preventive patrol experiment: A summary report. Report, National Policing Institute.
22. Klick, J. and Tabarrok, A. (2005). Using Terror Alert Levels to Estimate the Effect of Police on Crime. Technical report.
23. Kreps, D. M. and Wilson, R. (1982). Sequential equilibria. *Econometrica*, 50(4):863–894.

24. Lauri, M., Hsu, D., and Pajarinen, J. (2023). Partially observable markov decision processes in robotics a survey. *IEEE Transactions on Robotics*, 39(1):21–40. Publisher Copyright: IEEE.
25. Lei, X., Hu, X., Wang, G., and Luo, H. (2023). A multi-uav deployment method for border patrolling based on stackelberg game. *Journal of Systems Engineering and Electronics*, pages 99–116.
26. Lin, K. Y. (2022). Technical note–optimal patrol of a perimeter. *Operations Research*, 70(5):2860–2866.
27. Lin, M.-H., Tsai, J.-F., and Yu, C.-S. (2012). A review of deterministic optimization methods in engineering and management. *Mathematical Problems in Engineering*, 2012(1):756023.
28. Nagin, D. S. and Weisburd, D. (2013). Evidence and public policy. *Criminology & Public Policy*, 12(4):651–679.
29. National Academies of Sciences, E. and Medicine (2018a). *Proactive Policing: Effects on Crime and Communities*. The National Academies Press, Washington, DC.
30. National Academies of Sciences, E. and Medicine (2018b). *Racial Bias and Disparities in Proactive Policing*. The National Academies Press, Washington, DC.
31. Pawgasame, W. and Wipusitwarakun, K. (2021). Mobile sensor patrol path planning in partially observable border regions. *Applied Intelligence*, 51(8):5453–5473.
32. Portugal, D. and Rocha, R. P. (2016). Cooperative multi-robot patrol with bayesian learning. *Autonomous Robots*, 40(5):929–953.
33. Robin, C. and Lacroix, S. (2016). Multi-robot target detection and tracking: taxonomy and survey. *Autonomous Robots*, 40(4):729–760.
34. Royset, J. O. and Sato, H. (2010). Route optimization for multiple searchers. *Naval Research Logistics*, 57(8):701–717.
35. Samanta, S., Sen, G., and Ghosh, S. K. (2022). A literature review on police patrolling problems. *ANNALS OF OPERATIONS RESEARCH*, 316(2):1063–1106.
36. Schnelle, J. F., Kirchner Jr., R. E., Casey, J. D., Uselton Jr., P. H., and McNees, M. P. (1977). Patrol evaluation research: A multiple-baseline analysis of saturation police patrolling during day and night hours. *Journal of Applied Behavior Analysis*, 10(1):33–40.
37. Shafahi, A. and Haghani, A. (2015). Generalized maximum benefit multiple chinese postman problem. *Transportation Research, Part C: Emerging Technologies*, 55:261–72.
38. Shapiro, A., Dentcheva, D., and Ruszczynski, A. (2014). *Lectures on Stochastic Programming: Modeling and Theory, Second Edition*. Society for Industrial and Applied Mathematics.
39. Sherman, L. W., , and Weisburd, D. (1995). General deterrent effects of police patrol in crime "hot spots": A randomized, controlled trial. *Justice Quarterly*, 12(4):625–648. https://doi.org/10.1080/07418829500096221.
40. Sherman, L. W. (1997). *PREVENTING CRIME: WHAT WORKS, WHAT DOESN'T, WHAT'S PROMISING*, chapter Policing for Crime Prevention. U.S. Office of Justice Programs.
41. Sherman, L. W. and Eck, J. E. (2002). *Policing for crime prevention*. Taylor & Francis Group, Oxford, UNITED KINGDOM.
42. SHERMAN, L. W., GARTIN, P. R., and BUERGER, M. E. (1989). Hot spots of predatory crime: Routine activities and the criminology of place. *Criminology*, 27(1):27–56.
43. Smith, T. and Simmons, R. (2005). Point-based pomdp algorithms: improved analysis and implementation.
44. Spall, J. C. (2012). *Stochastic Optimization*, pages 173–201. Springer Berlin Heidelberg, Berlin, Heidelberg.
45. Szechtman, R., Kress, M., Kyle, L., and Cfir, D. (2008). Models of sensor operations for border surveillance. *Naval Research Logistics*, 55(1):27–41.
46. Telep, C. W., J., M. R., , and Weisburd, D. (2014). How much time should the police spend at crime hot spots? answers from a police agency directed randomized field trial in sacramento, california. *Justice Quarterly*, 31(5):905–933. https://doi.org/10.1080/07418825.2012.710645.

47. Wang, W., Dong, Z., An, B., and Jiang, Y. (2021). Toward efficient city-scale patrol planning using decomposition and grafting. *IEEE Transactions on Intelligent Transportation Systems*, 22(2):747–757.
48. Weisburg, D. (2008). Place-based policing. *Ideas in American Policing*, 9. Available: https://www.policefoundation.org/wp-content/uploads/2015/06/Weisburd-2008-Place-Based-Policing.pdf.
49. Yanes Luis, S., Peralta, F., Tapia Cordoba, A., Rodriguez Del Nozal, A., Toral Marin, S., and Gutierrez Reina, D. (2022). An evolutionary multi-objective path planning of a fleet of asvs for patrolling water resources. *Engineering Applications of Artificial Intelligence*.

Review Methodology

2

A systematic literature review procedure has been applied in this work to identify the most relevant and impactful studies on patrolling problems. Similar to many state-of-the-art surveys and reviews in the literature [5, 6], we first generated a search query to collect the research articles. We used the combination of keywords "patrol*", "optim*", "decis*", "security", "game", "military", "border", and "resource" to search through databases. At this stage, we also restricted the output to conference articles/proceedings and journal articles. We ran this query on three main databases (Compendex, Inspec, and GeoRef) via Engineering Village and the Web of Science Core Collection database. After removing duplicates, we were left with 1,132 articles/proceedings.

We then proceeded to the first screening phase by manually reviewing the titles and abstracts of these papers. The criteria at this stage were that the paper must clearly present a mathematical model related to patrolling, either directly or as one of the main outputs of the proposed model. As a result, 708 papers primarily focused on robotic improvements, with patrolling mentioned as an example in other domains, were removed. This left us with 424 journal articles and conference proceedings to examine.

At this stage, we also noticed that a significant number of promising conference articles/proceedings were either later published as journal articles within the same year or, at most, within two years of the conference publication (a trend mostly observed in OR/MS literature). Alternatively, some conference articles were later covered more generally in journal articles (a common pattern in computer science literature). To illustrate the first finding, [3] was followed by [4, 14] expanded and published as [15], and [8] was followed by [9] two years later. For the second finding, the features in [12] were summarized and expanded in [11], while the theories and application areas proposed in [1] were covered and extended in [2], similarly, the conference proceeding of [13] is expanded and published as [7]. These

Y. I. Tokel and J. Zhuang, *Optimizing Security Patrolling Strategies*, Synthesis Lectures on Operations Research and Applications, https://doi.org/10.1007/978-3-032-02617-0_2

observations led us to focus solely on journal articles, reducing the number of papers to examine from 424 to 216.

In the second screening phase, we thoroughly read 216 papers and assessed them based on their contribution to the literature, novelty of the approach, and whether their models could be classified under one of the following predefined themes: deterministic models, non-deterministic models, or game theory. After this detailed reading, we identified 111 papers. Additionally, through a snowballing approach, we discovered 11 more journal articles not found in the first screening, bringing the total to 122 articles at the end of the second screening phase. Figure 2.1 summarizes all the steps followed during the process [10].

Although we did not impose any restrictions on the publication year during our review, the resulting set of papers primarily spans the years 2006 to 2024, with a single paper

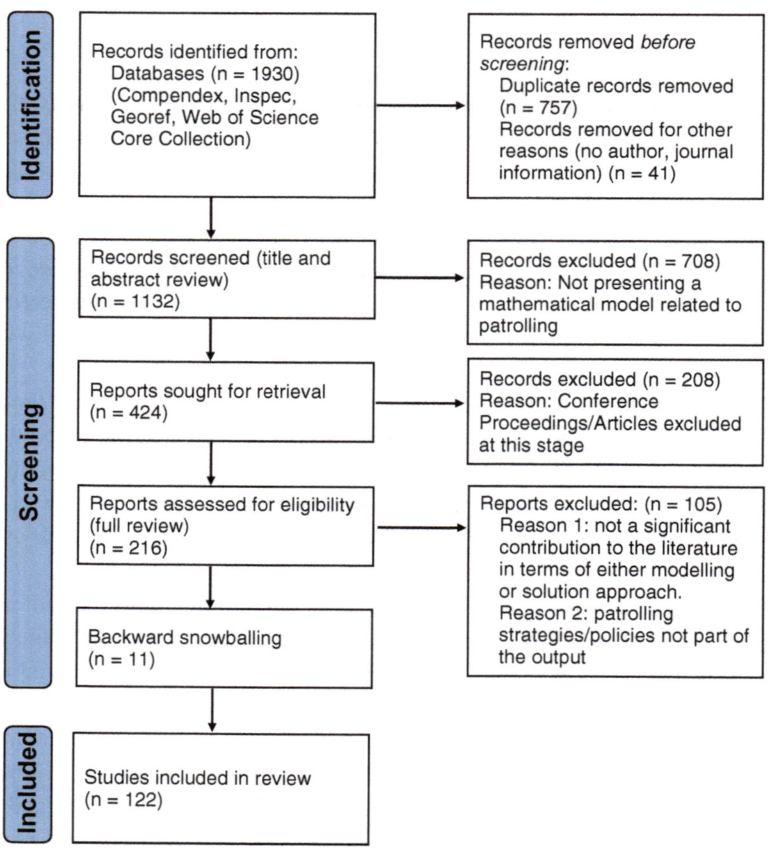

Fig. 2.1 PRISMA flow diagram illustrating the methodology of the systematic literature review

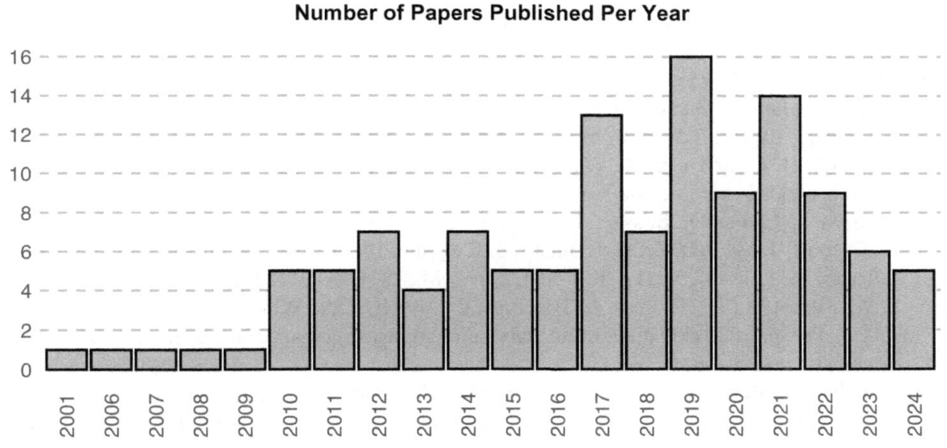

Fig. 2.2 Yearly distribution of the papers included in this study

published in 2001. Figure 2.2 illustrates the distribution of papers across publication years. It is worth noting that the majority of the papers were published after 2010, which might reflect a growing contribution of recent models to the field.

References

1. Basilico, N., Gatti, N., and Amigoni, F. (2009). Leader-follower strategies for robotic patrolling in environments with arbitrary topologies. In *8th International Joint Conference on Autonomous Agents and Multiagent Systems 2009, AAMAS 2009, May 10, 2009 - May 15, 2009*, volume 1 of *Proceedings of the International Joint Conference on Autonomous Agents and Multiagent Systems, AAMAS*, pages 30–37. International Foundation for Autonomous Agents and Multiagent Systems (IFAAMAS).
2. Basilico, N., Gatti, N., and Amigoni, F. (2012). Patrolling security games: Definition and algorithms for solving large instances with single patroller and single intruder. *Artificial Intelligence*, 184-185:78–123.
3. Guevara, C. and Santos, M. (2021). Crime prediction for patrol routes generation using machine learning. In *13th International Conference on Computational Intelligence in Security for Information Systems, CISIS 2020, September 16, 2020 - September 18, 2020*, volume 1267 AISC of *Advances in Intelligent Systems and Computing*, pages 97–107. Springer Science and Business Media Deutschland GmbH.
4. Guevara, C. and Santos, M. (2022). Smart patrolling based on spatial-temporal information using machine learning. *Mathematics*, 10(22):27.
5. Gupta, S., Starr, M. K., Zanjirani Farahani, R., and Ghodsi, M. M. (2020). Prevention of terrorism– an assessment of prior pom work and future potentials. *Production and Operations Management*, 29(7):1789–1815.
6. Hunt, K. and Zhuang, J. (2024). A review of attacker-defender games: Current state and paths forward. *European Journal of Operational Research*, 313(2):401–417.

7. Lin, E. S., Agmon, N., and Kraus, S. (2019). Multi-robot adversarial patrolling: Handling sequential attacks. *Artificial Intelligence*, 274:1–25.
8. McGrath, R. G. (2015). Patrol policies against strategic attackers at dispersed locations. In *IIE Annual Conference and Expo 2015, May 30, 2015 - June 2, 2015*, IIE Annual Conference and Expo 2015, pages 1185–1194. Institute of Industrial Engineers.
9. McGrath, R. G. and Lin, K. Y. (2017). Robust patrol strategies against attacks at dispersed heterogeneous locations. *International Journal of Operational Research*, 30(3):340–59.
10. Page, M. J., Moher, D., Bossuyt, P. M., Boutron, I., Hoffmann, T. C., Mulrow, C. D., Shamseer, L., Tetzlaff, J. M., Akl, E. A., Brennan, S. E., Chou, R., Glanville, J., Grimshaw, J. M., Hróbjartsson, A., Lalu, M. M., Li, T., Loder, E. W., Mayo-Wilson, E., McDonald, S., McGuinness, L. A., Stewart, L. A., Thomas, J., Tricco, A. C., Welch, V. A., Whiting, P., and McKenzie, J. E. (2021). Prisma 2020 explanation and elaboration: updated guidance and exemplars for reporting systematic reviews. *BMJ*, 372.
11. Paruchuri, P., Pearce, J. P., Marecki, J., Tambe, M., Ordonez, F., and Kraus, S. (2009). Coordinating randomized policies for increasing security of agent systems. *Information Technology & Management*, 10(1):67–79.
12. Pita, J., Jain, M., Marecki, J., Ordóñez, F., Portway, C., Tambe, M., Western, C., Paruchuri, P., and Kraus, S. (2008). Deployed armor protection: the application of a game theoretic model for security at the los angeles international airport. In *Proceedings of the 7th international joint conference on Autonomous agents and multiagent systems: industrial track*, pages 125–132.
13. Sless, E., Agmon, N., and Kraus, S. (2014). Multi-robot adversarial patrolling: facing coordinated attacks. In *Proceedings of the 2014 International Conference on Autonomous Agents and Multi-Agent Systems*, AAMAS '14, page 1093–1100, Richland, SC. International Foundation for Autonomous Agents and Multiagent Systems.
14. Trejo, K. K., Clempner, J. B., and Poznyak, A. S. (2016). Adapting strategies to dynamic environments in controllable stackelberg security games. In *55th IEEE Conference on Decision and Control, CDC 2016, December 12, 2016 - December 14, 2016*, 2016 IEEE 55th Conference on Decision and Control, CDC 2016, pages 5484–5489. Institute of Electrical and Electronics Engineers Inc.
15. Trejo, K. K., Clempner, J. B., and Poznyak, A. S. (2018). Adapting attackers and defenders patrolling strategies: A reinforcement learning approach for stackelberg security games. *Journal of Computer and System Sciences*, 95:35–54.

3 Analysis

In this section, we analyze our findings based on several categories, starting with the primary modeling themes. Our analysis revealed that the majority of research in the patrolling literature adopted Game Theory, Deterministic Optimization, or Stochastic optimization approaches, respectively as seen in Fig. 3.1a as their core modeling theme. Furthermore, we observed that almost all of the papers in the field utilized more than one modeling theme. To better represent the literature, we also identified secondary modeling themes, where applicable in Fig. 3.1b. The most common primary-secondary combinations included game theory with deterministic optimization, deterministic optimization with graph theory and game theory with either graph theory or stochastic optimization. It is also important to note that 10 papers that adopted deterministic optimization in our analysis did not adopt any secondary modeling approaches to either describe their problem or provide solution concepts and implemented everything within the scope of their primary modeling themes. Simulation also appeared to be a secondary modeling theme in both stochastic and deterministic optimization papers quite frequently.

The combination of game theory and deterministic optimization is the most commonly used, as researchers often model the environment as a game but rely on optimization techniques–such as MILP, LP, NLP, column generation, or genetic algorithms–to compute solutions or determine equilibrium points. Graph theory is also frequently employed, especially in deterministic or game-theoretic studies, to represent road networks as graphs more effectively. Stochastic optimization methods, including Markov decision processes (MDPs), partially observable MDPs, and Bayesian approaches, are also widely used by authors who frame their problems within a game-theoretic context.

Y. I. Tokel and J. Zhuang, *Optimizing Security Patrolling Strategies*, Synthesis Lectures on Operations Research and Applications, https://doi.org/10.1007/978-3-032-02617-0_3

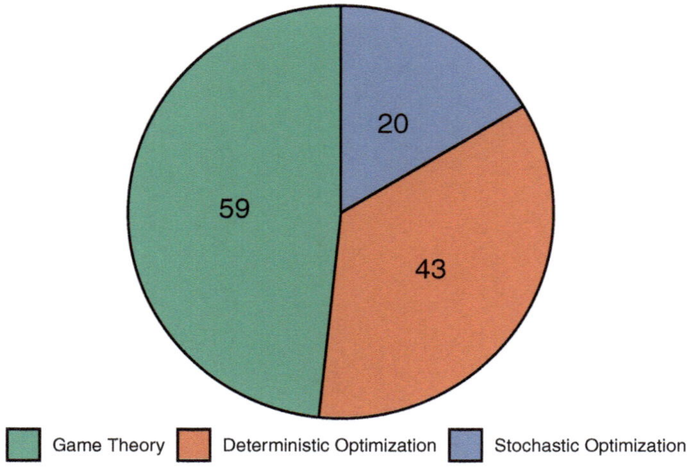

(a) Primary modeling themes

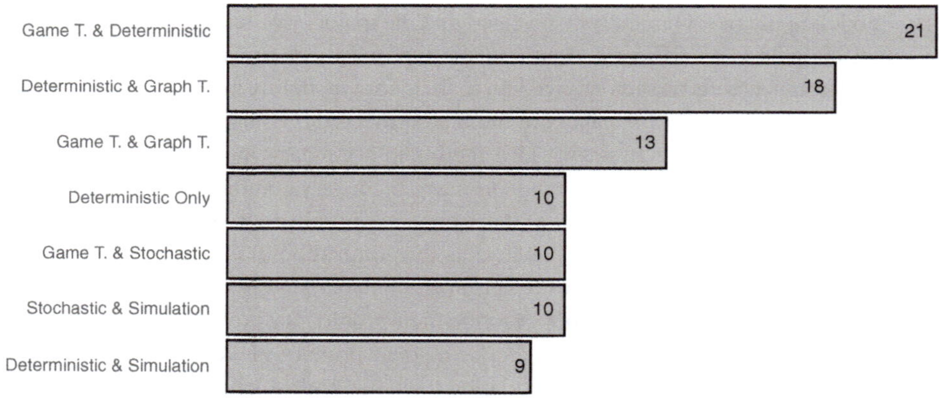

(b) Most frequent primary–secondary modeling theme combinations

Fig. 3.1 Distribution of primary and secondary modeling themes adopted in the literature

Another category we used in this work is to identify research papers according to the respective domains where they build or test their model as illustrated in Fig. 3.2. Note that although some papers provide motivating examples from relevant domains such as border or maritime security, if their results do not include any domain-specific implementation (e.g., [5–8]), we considered these as "Theoretic" during our analysis. Aside from this, the most frequently encountered domains in our analysis were urban, transportation, border, and maritime security. Of the papers that identified a specific domain in their work, the majority

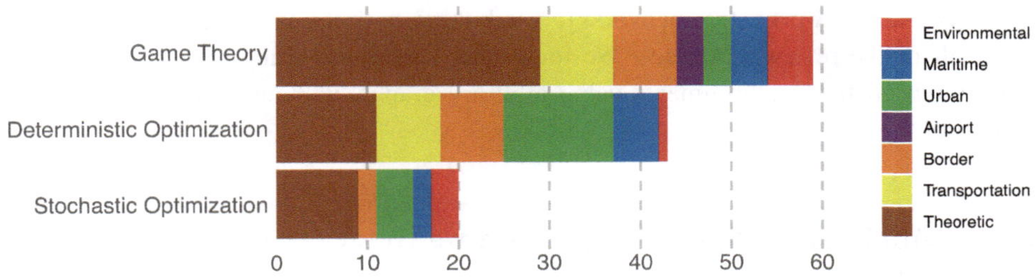

Fig. 3.2 Primary modeling themes with domain breakdown

provided realistic case studies. The remaining portion, while assigning a particular domain to their research, did not provide any real-world case scenarios. For example, in [53–55], the authors set up games between a customs officer and a smuggler but did not include any real instances in their work. Despite not implementing real cases, we categorized their domain as "Border Security" rather than "Theoretic," since the design, background information, and selected parameters were all related to border security. We observed that studies employing deterministic optimization approaches often presented more detailed and operationally grounded case scenarios compared to those utilizing stochastic optimization or game-theoretic models. One possible reason for this is that adding complexity to a deterministic model is relatively more manageable. In contrast, when randomness or strategic interactions are introduced, as in stochastic or game-theoretic models, the overall complexity increases significantly, often making the problem computationally intractable. Therefore, applying a deterministic model to a real-world scenario is typically less computationally expensive than using the other two approaches.

Upon examining the literature, we identified a crucial distinction in the focus of patrolling and categorized the papers accordingly as regular and adversarial, following the approach of [58]. In adversarial patrolling, the patroller's primary objective is to counter the opponent's actions, aiming to maximize the likelihood of capturing the adversary. In contrast, regular patrolling focuses on objectives such as increasing the frequency of visits (e.g., visiting key points or hotspots as often as possible) or maximizing the area covered [1]. Our analysis reveals that nearly all game-theoretic studies ($\sim$96%) focus on adversarial patrolling, whereas deterministic approaches almost entirely exclude such schemes, so that only 2 out of 41 papers employ adversarial patrol models. This disparity is unsurprising: deterministic formulations typically optimize a single, pre-specified objective and do not model the strategic, two-sided interplay between defender and attacker, whereas game theory is expressly designed to capture such bilateral interactions [108].

The representation of the area in which the problem is solved, the case is generated, or the game is played also varied, and in some instances, this was a key restriction. Following the approach in [12], we categorized the area of representation for the problem's realization as one of the following: open or closed perimeter, square or rectangular grid, special graphs (e.g., Eulerian networks, tree networks), or arbitrary, with the latter accounting for more than half of the reviewed articles. We found that papers providing realistic case studies often modeled their environments using arbitrary graphs, without specific graph-related constraints.

3.1 Analysis of Papers Adopted Game Theory as Primary Modeling Theme

During our analysis, Game Theoretic modeling was the most commonly adopted approach, motivating us to include additional metrics for a more thorough evaluation of these models. In addition to the categories summarized in Sect. 3.3, we introduced the following categories to enhance the analysis: information set (complete/incomplete), moves (simultaneous/sequential), period (single/multiple), second decision maker (i.e., Player 2; attacker), and solution concept (Nash, Stackelberg, Bayesian-Stackelberg, Perfect Bayesian).

In our analysis, only two research papers ([49, 50]) addressed the problem as a pure cooperative game, with the patrolling concentration modeled as regular. In [49], the authors introduced the Shapley function to measure the contributions and cooperation of mobile searchers (e.g., UAVs) in a search operation, where the objective was to minimize the time spent visiting all hotspots. In [50], the authors expanded on the model from [49] by incorporating a Petri net model to represent unreachable areas, with the goal of minimizing the length of sorties for mobile agents. The remaining papers focused on non-cooperative games, where the adversarial patrolling concentration was adopted.

Our analysis revealed that out of the papers we investigated, only [102] formulated their problem as 3 distinct players attacker, defender and informant(s). The informants in this game can choose to send a signal or remain silent, then if the defender receives a signal, defender updates her patrol strategy and allocate resources accordingly. The sequence of moves reads as the defender announces patrol strategy first, the attacker decides what target to attack, informant observes the attack intention perfectly and chooses to remain silent or inform the defender (which is private information), if the defender is informed she updates her patrolling strategy and attacker attacks to the target and outcomes are distributed.

In the following subsections, inspired by the methodology applied in [59], we will analyze the games based on their domain, information sets, and rationality. This approach will allow for a more detailed examination of the advantages and drawbacks of each game-theoretic model.

3.1.1 Model Domains

Almost half of the game-theoretic models analyzed did not belong to any particular domain as shown in Table 3.1; therefore we labeled them as "theoretic". Among these works, [5–8, 20, 43, 45, 49, 78, 133] and [50] were able to provide analytical solutions to their models. However, some of their solutions were restricted to specific area representations. For example, [6, 8, 78], and [133] only provided solutions for closed perimeters, while [7] solved the game for open perimeters (i.e., line graphs) and for certain arbitrary graphs under specific conditions (e.g., even periods). Alpern et al. [5, 20], and [45] offered analytical solutions for special graphs. Specifically, [45] provided optimal patrolling strategies for Eulerian networks, while [5] offered solutions for (i) any network with short attack times, (ii) tree networks that satisfy specific conditions, and (iii) star networks with one long arc and equal-length others. Later, [20] extended this model, proving that the mixed strategy for the patroller (E-patrolling strategy) is optimal for all tree networks and solving a special type of complete networks. Fu et al. [43] provided analytical solutions for a differential game, focusing on optimal strategies for a slower defender and attacker in a small target area. Other "theoretic" game-theoretic studies adopted mathematical optimization methods, such as linear programming [48, 73, 80, 83, 85], mixed-integer linear programming [90], [23, 65], and/or non-linear programming [35, 47, 105, 110, 113]. Additionally, some used stochastic modeling approaches, such as Markov decision processes [35, 47, 105, 110, 111], or partially observable Markov decision processes [36, 113]. Most of these studies also employed heuristic algorithms [48, 79, 80, 85, 113] to generate approximate solutions, especially when the problem size was large.

Eight of the works were labeled as "transportation security" during our analysis, and six of them provided realistic case studies. In [126], the authors modeled a patrolling problem

Table 3.1 Domain-specific classification of studies that adopt game-theoretic models and whether they are supported by a realistic case study

Domain	Realistic case study	Papers
Theoretical	Provided	[92, 125]
	Not provided	[5–8, 12, 20, 35, 36, 43, 45, 47–50, 65, 78–80, 83, 85, 90, 102, 105, 110, 111, 113, 133]
Airport security	Provided	[4, 60, 61]
	Not provided	–
Border security	Provided	[17]
	Not provided	[54, 54, 55, 69, 73, 89]
Environmental security	Provided	[24–26]
	Not provided	[96, 128]
Maritime security	Provided	[16, 18, 88]
	Not provided	[9]
Transportation security	Provided	[15, 37, 38, 40, 126, 127]
	Not Provided	[13, 23]
Urban Security	Provided	[11, 19, 52]
	Not Provided	–

between a defender and an attacker, where they simultaneously play the game. The defender chooses a schedule to minimize the damage, while the attacker chooses a time and target for the attack. Due to the large size of the problem, where they simulated a real rail network in Philadelphia, they developed a column generation procedure to solve the game and used dynamic programming to solve the subproblem. Later, they extended this model in [127] to introduce more than one defender and attacker, time-dependent node values, and node-specified attack times. The solution approach involved developing column and row generation algorithms to solve the problem. They used the same transportation rail instance as in [126] with multiple patrollers and concluded that as more patrollers are added to the network, the reduction rate slowly decreases with each additional patroller (diminishing marginal return effect). In [15], the authors adopted a Stackelberg game approach to model interactions between the transit operator and opportunistic passengers. They precisely calculated the inspection probability with a function and developed heuristics to solve large-scale problems within a reasonable time. As a realistic use case, they conducted their experiments on the LA Metro Rail line to generate optimal, unpredictable patrolling schedules. Similarly, in [37], the authors presented a Bayesian Stackelberg game model with stochastic uncertainty, using Markov decision processes to study the interaction between fare evaders and transit operators. They used the LA Metro Rail as the testbed for their studies and compared their approach with human-generated patrolling schedules. They reported that the schedules generated by the game model outperformed other models in terms of the number of detected fare evasion incidents. Later, in [38], they improved their models and introduced branch-and-price-based algorithms called $SMART_0$ (exact algorithm) and $SMART_H$ (heuristic) to better allocate defender resources (i.e., patrollers) to maximize utility. They conducted a large-scale real-world experiment with 80 security officers patrolling 10 stations for one day. Again, compared to human-generated schedules (which are the current procedure), their schedules improved coordination and reduced workload by automating the schedule-generation process. Along similar lines, [40] proposed a single-level linear optimization problem to solve the fare inspection patrolling strategy and unpredictable patrolling schedule problems. They adopted a column-generation approach, with the related subproblems as outputs: the pricing subproblem generated the schedules, and the master problem identified the probability distribution among the potential patrolling schedules to be selected. They simulated the environment using Monte Carlo simulation and achieved a 2.2% evasion rate after 120 days, concluding that their methodology for unpredictable patrolling schedules could be easily and efficiently adapted to real-world scenarios.

We identified seven papers related to the border security domain, of which two provided realistic case studies. The papers that did not provide any use-case scenarios adopted the following solution methodologies: (i) analytical solutions [89], partly [53], (ii) deterministic optimization methods, such as LP [54, 55], dynamic programming [53], (iii) stochastic methods, such as Markov decision processes [53], or (iv) column generation [69]. In [17], the paper provides three generic examples related to protecting critical infrastructures, one of which is the border patrol problem [94]. They studied a resource-allocation model to optimally allocate resources (road patrol, helicopter, remote observation post) with a limited budget to minimize the maximum probability of nondetection. Finally, in the pursuit-evasion game [70], a single pursuer tries to capture multiple trespassers before they reach the safe house, assuming the patroller is faster than the evader. To find optimal solutions, they developed a novel rapidly-exploring random trees algorithm for obstacle avoidance and pathfinding, called SP2E and SP3E (for solving problems with 2 and 3 evaders, respectively). They found the minimax solution for the one patroller-one evader case. They first validated their results through simulation and then by programming robots, achieving similar results.

We identified five papers that developed their models in the domain we generalized as "environmental security". Of these, three provided realistic case studies. In [24], a Stackelberg game is proposed for a chemical cluster, where the scheduling of patrol resources is modeled as one of the players against multiple attackers characterized by illegal releases. They implemented their proposed solution (using multiple LP methods) in a chemical cluster in Shanghai and concluded that their solution provided a much higher probability of detecting the attacker's illegal releases compared to randomized or fixed patrolling schedules. In [26], case studies were conducted in Shanghai, Antwerp, and Houston chemical clusters, using real data for patrolling scheduling operations. First, they partitioned the chemical clusters based on two principles: "distance first" and "balancing." Based on this partition, they solved the game for different patrolling cars in each partition. The results of the Stackelberg equilibrium demonstrated that by adopting a mixed Stackelberg strategy, the defender could improve their utility by 20% and reduce the attacker's effectiveness by more than 30%. In a follow-up study, [25] further improved their model by introducing bounded rationality of the attackers and the uncertainty of the defender's estimation of the attacker's parameters. They implemented their new solution method in the chemical cluster in Antwerp, which was used in [26]. Although the payoffs were slightly reduced for the defender after adding two layers of uncertainty, this made the model more robust and ready for implementation in a real-world scenario. The remaining two studies, [128] and [96], implemented Stackelberg equilibrium models. The former incorporated a robust solution to address scenarios with incomplete information about the attacker, while the latter adopted a risk-based framework to assess security threats to the pipeline system under consideration. To optimize the route, [128] estimated upper and lower bounds of the data and employed the Multi-LP method to determine the defender's optimal fixed route. They reported that the performance of proposed approach outperformed pure randomized as well as fixed patrolling routes.

Next, we identified four papers in the maritime security domain, three of which provided realistic case scenarios. In [16], a realistic attacker-defender game is modeled between multiple attackers (speedboats) and multiple defenders (patrol boats), where the defender tries to minimize the maximum probability of evasion by the attackers. To solve the problem, they first developed a mixed-integer linear program, followed by a minimax algorithm. The area representation of the game is a rectangular grid, and a sample scenario based on the US Navcent Fifth Fleet in Bahrain and the Port of Hong Kong was used to test the algorithm and modeling effort. In [18], illegal fishing in the Gulf of Mexico was selected as the use case to test their MIDAS algorithm, which takes into account both bounded rationality and incomplete information to generate optimal patrol routes for the patrollers. In [88], the authors selected the Gulf of Aden to validate their model using a custom-built simulation tool. Their findings revealed that when the time horizon is too long, prioritizing patrolling strategies that maximize presence becomes more effective than those based on anticipating the adversary's movements. Although there was no real-case scenario studied in [9], the motivation and examples are all drawn from maritime security, and they modeled both non-strategic (e.g., probabilistic) and strategic (e.g., perfect rationality) cases. The strategic case is modeled as a Blotto game, where two players distribute their resources simultaneously on a grid. The scenarios with one interceptor and one interdictor, as well as multiple interceptors and one interdictor, are studied, and analytical solutions are provided.

We identified the relative domain of three papers as "urban security," and all of them provided realistic case studies. In [11], their model is tested in a fair site simulation with 700 security guards, 170 sensitive locations, and 130 targets, of which 35 are identified as critical. Based on signals generated from sensors, they optimized response strategies so that a security force could reach the area where the alarm was initiated in the minimum time. In [19], the authors studied a Stackelberg game in which the solution of the game would be a schedule for night patrols. The scenario is implemented and illustrated by imitating the Arica Parinacota Region of Chile. The defender (Chilean national police force) deals with three types of crimes, for which they know the probabilities of occurrence: drug trafficking, contraband of goods or animals, and illegal entry. They provide a formulation (COMP) and implement branch-and-cut algorithms to generate an optimal solution. Finally, [52] proposed a game model where the attacker attacks edges, and patrollers conduct surveillance by patrolling the network with multiple vehicles. The defender's objective is to intercept the attacker by traversing the edge during the attacker's attack. To solve this mixed-integer problem, they used a decomposition method faster than the column generation algorithm. To test their approach, they built a realistic instance projecting downtown Santiago: a graph with 119 nodes and 203 edges; however, they admitted that they would need a better use case to test the algorithm.

The last category in our domain-oriented categorization is "airport security." We identified 3 papers that fall into this category, and all provided real-world scenarios during their analysis. In [4], they modeled a partially observable Markov decision process to address the probability of inaccuracy during the patrolling process. They provided a numerical example to illustrate their model's performance in a scenario involving airport terminals. Janssen et al. [61] developed a model to address the problem of subjective payoff estimation by experts using agent-based modeling. They performed a case study and simulated a regional airport where one patroller protects 4 targets from an incoming attack, and attackers choose one target to attack. Both players are utility maximizers and assumed to be perfectly rational. Their results showed that by strategically randomizing patrol routes, higher expected rewards for the security officer were achieved. Finally, in [60], they summarized the ARMOR and IRIS scheduling assistants, which were actually deployed and used by the Los Angeles Airport

(LAX) Police and the Federal Air Marshal Service (FAMS), respectively. They provided effective algorithms to solve large-scale Bayesian-Stackelberg games to generate efficient patrolling schedules in an adversarial environment.

3.1.2 Information Sets

While playing a game, a player's information set regarding their opponent can either be complete or incomplete. In the case of complete information, the player has full knowledge of all potential actions their opponent could take. However, information may sometimes be hidden, either due to imperfection or incompleteness. Imperfection refers to situations where the observed information is imperfectly perceived (e.g., a sensor failing to capture the precise state of an event). Incompleteness, on the other hand, occurs when some information is deliberately withheld from the player (a concept often explored in signaling theory, for applications of signaling theory in homeland security, see [131, 132], and [109]). Although these concepts are formally separate—with imperfection relating to observability and incompleteness to knowledge asymmetry—in this work we simplify our classification by treating any deviation from perfect or complete information as a case of "incomplete information." This includes games with hidden types, unobservable moves, partial observability, or belief-based reasoning. Thus, in our review, we categorize all such models as incomplete information.

As shown clearly in Table 3.2, the Bayesian-Stackelberg equilibrium is the most commonly adopted concept within incomplete information structure among these papers. In all [18, 19, 26, 37, 60, 90, 92] and [96] there are different types of attackers, and the patroller (she) individually knows their payoff structure and other related parameters; however, during the realization of the game, she does not know which type of adversary she will face, and the incomplete information structure comes into play. As the first mover, the patroller must plan before patrolling to be prepared for different types of attackers.

Basilico et al. [12] proposed Patrolling Security Games (PSG) as two-player, multi-stage games with imperfect information. Each turn, both the patroller and the attacker play simultaneously, and the patroller cannot directly observe the attacker's moves (e.g., the attacker's location is not perfectly observable). They developed efficient algorithms to compute leader-follower games on large scales. Following, [11] proposed a new approach to support patrollers with a wide range of strategies. However, these sensors are not "perfect," so after receiving a signal, the patroller is unsure of the exact location of the incident but can approximate the area. They provided novel algorithms to solve the game and proved that the best the patroller can do (assuming sensors do not produce false positives or miss any detections) is to stay idle at one position before receiving the signal and respond only after the signal is received. They provided a large-scale real case study to validate their algorithms and strategies. Hohzaki and Masuda [55] proposed a zero-sum game between a patroller and a smuggler, where the smuggler tries to smuggle goods to the border and the patroller tries to prevent it. In this work, the amount of contraband the smuggler carries is private information, and the patroller has a probabilistic belief about the quantity. Therefore, a PBE

Table 3.2 Equilibrium concepts in non-cooperative games, with corresponding information structures, move types, and model types

Equilibrium concept	Information type	Moves	Papers
Nash (NE)	Complete	Simultaneous	[6–9, 13] (static infrastructure model), [20, 54, 79, 80, 85, 89, 113, 126, 127, 133]
Nash (NE)*	Complete	Sequential*	[5, 20] [43, 45, 78]
Perfect bayesian (PBE)	Incomplete	Sequential	[12, 54, 102]†
Stackelberg (SE)	Complete	Sequential	[15, 17, 17, 23, 24, 26, 38, 40, 48, 52, 61, 65, 73] [88, 125, 128]
Stackelberg (SE)**	Incomplete	Sequential	[4, 11, 35, 47, 69, 110]
Bayesian stackelberg (BSE)	Incomplete	Sequential	[18, 19, 25, 37, 60, 90, 92, 96]
Non-standard equilibrium approach	Incomplete	Sequential	[13] (POMDP-based equilibrium strategy for dynamic game), [36] (POMDP/POMG-based equilibrium strategy), [83] (A new equilibrium concept called ACMR)

* *Papers in this category adopt a continuous-time game structure and employ Nash equilibrium as the solution concept. Although the players' actions unfold over time, the game does not involve classic turn-based sequential moves with observable commitments (as in Stackelberg or extensive-form games). Instead, strategies are planned over a continuous horizon where both players optimize simultaneously in a dynamic setting. We therefore label them as "sequential" in a broad temporal sense, but emphasize that the equilibrium is still of the Nash type and not a refinement like SPNE or PBE*

** *Papers in this category are modeled as classic Stackelberg games, where one player (the leader) commits to a strategy first and the other (the follower) best-responds, with both players having full knowledge of the game structure. However, these models incorporate incomplete information–either through uncertainty about specific parameters (e.g., spatial uncertainty), imperfect observations available to one of the players, or through adaptive mechanisms such as reinforcement learning where players gradually improve their strategies over time based on partial feedback. Since we treat imperfect information as a form of incomplete information in this study, such games are categorized under Stackelberg equilibrium with incomplete information*

† *Shen et al.* [102] *introduces a three-player Stackelberg-perfect Bayesian equilibrium. Although it includes an additional player (strategic informant), the equilibrium still fits within the PBE framework due to its consistent belief-updating and sequential rationality structure*

was adopted as a solution concept, and a numerical LP was developed to approximate the results. Finally, the study by [102] adopts a hybrid framework that integrates a Stackelberg game between the defender and the attacker with a PBE structure between the defender and strategic informants. To determine the optimal defensive strategy, the authors solve an LP for each target and select the most effective plan from the resulting solutions.

During our analysis, we identified a group of studies that incorporate imperfection, incompleteness, or uncertainty within the game environment, yet do not apply Bayesian belief structures to handle these uncertainties. We classify these studies under Stackelberg Equilibrium with incomplete information, as they feature uncertainty either in the environment, actions, or observations, without relying on Bayesian-type modeling. For instance, [4] address uncertainty regarding each player's knowledge of the opponent's state by modeling the environment as a partially observable Markov decision process (POMDP). They develop a two-step NLP-based algorithm that regularizes a penalty function to handle incomplete information. In [35], although the attacker observes the defender's policy, the environment itself evolves with probabilistic transitions and uncertain state changes. The model incorporates belief state dynamics and partial observability, making it a sequential game with incomplete information. Similarly, [47] introduce entropy-based reasoning to model epistemic uncertainty about target conditions–such as unknown states or locations of attackers–without employing any Bayesian framework. The defender's patrol strategy is shaped by this inherent uncertainty in the game's realization. Laan et al. [69] propose a sensor-aided patrolling model as a partially observable agent-intruder stochastic game, where sensors detect intruders probabilistically. The authors provide Nash equilibria for finite POAIGs and approximate (ϵ-optimal) solutions for infinite-horizon cases. In [110], while players have full knowledge of the game structure, the uncertainty arises from probabilistic transitions and lack of mutual observability during execution. This uncertainty is execution-based rather than due to hidden types or strategic asymmetries–hence, it is not treated through Bayesian updating. Lastly, in [11], the source of imperfection lies in spatial alarm uncertainty. Although both players know the game rules, the defender receives imprecise signals from wide-area sensors, which obscure the exact location of attacks and force the defender to act under incomplete spatial information.

We also identified three studies that deviate from classical equilibrium formulations. These were collectively categorized under Non-standard equilibrium approaches in Table 3.2. Baykal-Gursoy et al. [13] modeled a game where the patroller cannot observe the immobile adversary's location; they proposed a POMDP model and developed dynamic programming algorithms to solve the game. As an output of the game, the first responder (patroller) can obtain the optimal patrol strategy to minimize (or maximize) the total expected risk (or reward). In [36], the authors designed a Bayesian-Markov game. The imperfection was observed in the states of the partially observable Markov decision process (POMDP). Reinforcement learning was adopted to estimate transition matrices for each agent at each time step through simulations. The unique contribution of the paper lies in their approach to combine three fields (game theory, design theory, and reinforcement learning) to create a novel nucleus. Finally, [83] introduced a new regret-based equilibrium ACMR that does not conform to classical equilibrium concepts like Nash or Stackelberg, but still models adversarial interactions with incomplete attacker observability.

The information structure does indeed complicate the results significantly. However, if the goal is to model problems as realistically as possible, the assumption of complete information must be relaxed more frequently as stated in [83] and [59]. In modeling environments to achieve optimum surveillance or patrolling, the complete information assumption often makes models less applicable to real-world scenarios. Therefore, in this subsection, we aimed to highlight existing approaches to relax this assumption and better align the models with practical applications.

3.1.3 Rationality

Perfect rationality of agents is a common assumption in many game-theoretical studies. According to this assumption, both players play the game strictly to maximize their own benefit, reward, or utility. However, in real-life scenarios, it is known that actors might not always be perfectly rational [103]. For example, a smuggler observing a perimeter to infiltrate through a border might not execute his plan perfectly if he encounters something unexpected. Such uncertainties are quite common in real-life situations. To reflect this in modeling efforts, many researchers explore ways to relax the perfect rationality assumption. One widely accepted approach is the notion of bounded rationality. In our study, however, we only identified four papers in which the perfect rationality assumption was relaxed.

In [92], the authors challenged prior models that assumed players were pure utility maximizers. They introduced a new algorithm, COBRA, that incorporated bounded rationality and anchoring bias parameters. To test the effectiveness of this algorithm, they conducted a human subject experiment with 218 participants, comparing the results of human decisions with those predicted by the new algorithm and previous models assuming perfectly rational agents. They found that the new model provided statistically significant results, closely matching the human subjects' decisions, and concluded that bounded rational models better reflect real human decision-making processes.

Following this, [125] expanded on this model by introducing Prospect Theory ([63]) and Quantal Response Equilibrium to account for risk preferences, such as loss aversion, and variations in how humans interpret quantitative assignments. Their results showed that these improvements led to statistically significant better outcomes compared to the benchmark algorithms, including the original COBRA model. Later, [18] proposed the MIDAS algorithm, which uses a robust maximin formulation to handle uncertainty-induced decision-making processes. They replaced the previous quantal response model with a new subjective utility quantal response model (SUQR). By combining these two features, MIDAS addresses both scalability and robustness, producing patrol schedules that consider boundedly rational adversaries.

In a similar vein, [25] relaxed the perfect rationality assumption in their previous study [26] involving the area-partition collaborative patrolling (APCP) game. They implemented an ϵ-optimal response strategy developed in [92], allowing for attackers who may not be strictly optimal but are instead off by an ϵ amount. Additionally, they changed some parameter values to distributions to better model the uncertainty involved in real-life patrolling scenarios, validating their results through simulations.

This line of research shows great promise and is likely to see increasing attention. However, due to the significant computational costs involved in developing such models, extensive fine-tuning is required, which has limited its widespread adoption. This is partly why we identified only four relevant studies during our analysis. Despite this, there are numerous opportunities to apply these approaches in the patrolling domain, and we expect further developments in this area will enhance its practical implementation.

3.2 Analysis of Papers Adopted Deterministic Optimization as Primary Modeling Theme

In this section, we examine papers where deterministic optimization methods are the primary focus. The most frequently used solution methodologies include linear programming [72, 129], mixed integer (linear) programming (MIP/MILP) [21, 22, 30–32, 32, 33, 51, 56, 66, 71, 84],[67, 76, 100, 112, 124, 130], integer (linear) programming (IP/ILP) [27, 57, 114], (mixed integer) non-linear programming (MINLP/NLP) [2, 29, 34, 97, 99], and binary (integer) programming (BP/BIP) [10, 62, 81, 104, 118].

In both [34] and [97], although the evader or target is not explicitly modeled as a rational player, the primary objective is to optimize detection: maximizing the probability of detecting the evader in a maritime context in the former, and minimizing the probability of not detecting the target in the latter.

Most of the papers analyzed in this category did not impose any specific network or graph structure as a constraint for their models. However, [29] and [74] considered a rectangular (square) area to model their problem. Since they did not discretize the area into smaller segments (unlike [34, 81, 97, 123, 129] where the area was discretized into a square, or hexagonal segments), any point in the area can become a target. Given that their focus is on maritime security, they argued that this assumption did not limit the general applicability of their model. Seon Jin et al. [99], and later [2], modeled an open perimeter as the representation of the area for deploying wireless charging systems in a border security context. On the other hand, [32] and [56] only provided solutions for closed perimeter areas, and since their work is primarily theoretical, the impact of this limitation is difficult to assess.

3.2.1 Model Domains

In this section, we categorized the reviewed studies based on their application domains and the presence of realistic case studies to better understand how deterministic optimization models are applied across various operational settings. This classification helps reveal both the breadth of practical use and the areas where research remains primarily theoretical. Table 3.3 organizes the selected papers into five main domains: Theoretical, Border Security, Maritime Security, Transportation Security, and Urban Security, and distinguishes whether each study grounds its model in real-world data.

Approximately a quarter of the reviewed papers did not focus on a specific operational domain and were therefore categorized as theoretical in our analysis. These studies primarily aimed to develop general-purpose methodologies or algorithms, often validated through synthetic or simulated environments. Royset et al. [97] modeled a generalized search problem involving multiple searchers and probabilistically moving, non-strategic targets. Although they mentioned possible applications in military, rescue, border patrol, and law enforcement contexts, the absence of a concrete case study and the model's generalizability led us to classify it as theoretical. Similarly, [32] proposed methodologies for identifying optimal circulatory loops for patrolling vehicles, focusing on the strategic selection of origin locations under different allocation strategies, but without tying the model to a particular domain.

Some of the studies focused on multi-robot or UAV-based patrolling problems. In [121], the authors developed a distributed algorithm based on expected idleness to improve patrolling effectiveness in cooperative multi-robot settings, validated through both sim-

Table 3.3 Domain-specific classification of studies that adopt deterministic optimization models and whether they are supported by a realistic case study

Domain	Realistic case study	Papers
Theoretical	Provided	–
	Not provided	[27, 30, 32, 33, 41, 56, 62, 84, 95, 97, 121]
Border security	Provided	[2, 22, 66, 70, 81]
	Not provided	[86, 99]
Environmental security	Provided	–
	Not provided	[123]
Maritime security	Provided	[29, 57, 129]
	Not provided	[34, 74]
Transportation security	Provided	[10, 21, 51, 71, 104, 130]
	Not provided	[118]
Urban security	Provided	[31, 42, 67, 72, 75, 76, 87, 100, 112, 114, 116, 124]
	Not provided	–

ulated environments and simple corridor tests with real robots. Farinelli et al. [41] also addressed multi-robot patrolling, emphasizing edge interferences in the environment–an aspect often overlooked in node-focused formulations–and validated their coordination framework using simulators and corridor-based robotic experiments. Expanding on this direction, [84] explored patrolling with heterogeneous robots and assessed their method using a custom-built synthetic simulation platform. In contrast, [30] and [95] focused on UAV task assignment with rechargeability and inter-vehicle cooperation, respectively, and tested their algorithms in various simulated network environments without anchoring them in specific real-world scenarios. Jing et al. [62] studied UAV sweep coverage and also relied on controlled simulation environments for evaluation.

Several theoretical studies tackled routing and scheduling formulations using advanced optimization tools. Chen [27] introduced an approximate cross entropy algorithm for patrol route planning in dynamic environments, demonstrating its computational efficiency through simulation. Chircop et al. [33] presented an integrated routing and scheduling model solved via a branch-and-price framework enhanced by problem reduction techniques; while some networks were inspired by real-world settings, the data remained simulated. Finally, [56] addressed the Length-Constrained Cycle Partition Problem (LCCP), proposing both exact MIP formulations and a heuristic algorithm. Their approach, tested on TSPLIB-based instances with synthetic constraints, produced high-quality solutions efficiently, with conflict hypergraph-based constraints improving model performance. In the border security domain, we found that most studies provided realistic case studies to validate their proposed models, with the exceptions of [86] and [99]. Lau et al. [70] formulated a pursuit-evasion problem in which a single unmanned vehicle aims to intercept multiple non-strategic intruders. Beyond simulation, they validated their online path planning algorithm through experiments with real unmanned ground vehicles in a controlled environment. In contrast, [99] developed a drone-aided border surveillance system that incorporated an electrification line (E-line) for in-flight drone recharging. While they introduced optimization models to allocate E-line segments and minimize the required number of drones, their validation relied solely on simulated data. Building on this work, [2] extended the model to incorporate additional features such as surveillance priorities for different locations and variable drone speeds based on coverage demands. Importantly, they supported their approach with a realistic case study involving a 22.8-mile segment of the U.S.–Mexico border, demonstrating the practical feasibility of continuous drone surveillance with E-line support. Similarly, [66] conducted a case study on the Türkiye–Syria border, dividing the 910-km boundary into 152 checkpoints and four depots to evaluate the performance of their matheuristic approach under operational constraints. Liu et al. [81] used the China–Vietnam border segment as a motivating example to validate their algorithms. Although the data was not directly collected from the field, the model was explicitly tailored to the region's characteristics, and practical implications were discussed–thus qualifying it as a realistic case study. Cardoso et al. [22] focused on resource allocation for border surveillance using a black hole-based metaheuristic. To validate the model's effectiveness, they applied it to five different real-world instances along the

Brazilian border, incorporating actual surveillance schedules and hotspot data. Lastly, [86] framed their optimization model within a border security context and discussed motivational examples; however, they relied on a fictional setting–referred to as the Boulevard Region–to carry out simulation experiments.

The only paper in our review that modeled a problem in the domain of environmental security under the deterministic optimization category is [123]. The authors used a discretized representation of the Ypacaraí Lake in Paraguay to define the operational environment for simulation. While the use of a real-world geographic region adds contextual relevance, the study did not incorporate actual environmental data from that location. Instead, values for water quality parameters such as pH, temperature, conductivity, and turbidity were assumed and treated as if they had been obtained in advance through informative path planning techniques. As the model inputs and validation remain entirely synthetic, we categorized this work as not containing a realistic case study despite its geographically grounded scenario.

Out of the five studies in our review categorized under the maritime security domain, three provided realistic case studies to validate their approaches: [57, 129], and [29]. In [57], the authors developed an optimization-based software toolkit to support the scheduling of patrol boats and personnel for the Royal Australian Navy. To test their models, they conducted scenario-based simulations using estimated data derived from actual operational constraints. Similarly, [129] proposed a risk-based path planning framework to support rescue ship patrolling at sea, utilizing publicly available data from various sources related to the Northeastern South China Sea. They also collaborated with five marine experts to validate risk index values, acknowledging the limitations of data granularity while emphasizing that further precision was constrained by cost and technical feasibility. Chen et al. [29] developed a patrol path planning model that incorporates Automatic Identification System (AIS) data and Aerial Photographs (APs). They evaluated their approach through three scenarios: one without AIS/AP data, one with both integrated, and a third combining elements of the first two. The study was conducted using two illustrative instances–a small and a relatively large area–based on the Gulf of Guinea, a region identified as a maritime crime hotspot. The authors estimated model parameters through a comprehensive review of publicly available data on maritime crimes. In contrast, [34] presented an optimal control framework for maximizing the detection rate of intruders in a channel patrolled by unmanned underwater vehicles. Although their model was well-structured, its validation was limited to purely simulated environments across single-, two-, and three-patroller cases. Similarly, [74] addressed a maritime rescue scenario under severe weather conditions using a biologically inspired two-stage metaheuristic. Their algorithm was tested extensively in a simulation environment, but without reference to a specific operational setting or real-world data.

Our analysis identified seven studies that focused on the transportation security domain. Among them, six incorporated realistic case studies, while one relied solely on synthetic data. Zhu et al. [130] developed an optimization toolkit to evaluate and compare patrolling versus prepositioning strategies for emergency traffic response units. They validated their model using empirical data from the I-495/I-95 Capital Beltway, drawing on incident records spanning 2006 to 2010. Similarly, [71] addressed the scheduling of security patrol units across a mass rapid transit system and used the Singapore MRT rail network as a case study. In [10], the authors proposed a binary programming model aimed at maximizing vehicle flow coverage on an interurban road network during a fixed time horizon. To evaluate their approach and supporting heuristics, they applied it to a realistic network based on an urban city in Northern Israel. In [21] authors presented an improved formulation of the Maximum Covering and Patrol Routing Problem (MCPRP) to optimize highway patrol routes for state troopers using hotspot data on crash-prone locations and times (rural areas of Jefferson, Mobile and Tuscaloosa counties in Alabama). By strengthening the original model and reducing solution space, they reported that the new formulation achieved significantly better hotspot coverage while solving large instances more efficiently. Snijders et al. [104] addressed the scheduling of security teams for the Dutch railway network, aiming to shift from decentralized, low-detail planning to a centralized, detailed scheduling process to enhance social safety. The authors model the problem as a shortest path and set covering problem with added constraints, adapting and extending an existing Lagrangian-based heuristic to fit the unique requirements of security operations. They implement and test their solution through a software prototype, using real data provided by the Netherlands Railways. He et al. [51] introduced a novel model where a patrol vehicle and a drone coordinate to complete heterogeneous road-based tasks in an urban environment. The authors modeled the problem as a double-layer arc routing problem using a two-layer mixed-integer programming formulation and propose a two-stage heuristic algorithm. A case study on Hefei, China road network demonstrated the algorithm's effectiveness in producing efficient, real-world traffic patrol solutions. In contrast, [118] did not incorporate any real-world case study. Instead, they developed a coordinated route planning algorithm for a vehicle-drone traffic patrol problem and tested their model using synthetically generated networks of varying sizes.

Urban security emerged as the most represented domain within the deterministic optimization category, with 12 studies, all grounded in realistic case studies. Several studies focused on optimizing security patrol operations in residential and urban environments. Lau et al. [72] tackled the problem of persistent resident complaints in large estates in South China despite increasing patrol staff. They proposed an intelligent deployment system that optimized guard allocation using complaint data to minimize future incidents. Similarly, [116] studied patrol route optimization in Gauteng, South Africa, introducing a tabu search algorithm that achieved more balanced work distribution than existing practices. Vidigal Correa et al. [112] addressed a complex multi-period orienteering problem for Coopservice, a major Italian firm, developing a MILP and an Iterated Local Search (ILS) heuristic with variable neighborhood descent (VND). Tested on 79 real-world instances across 12 provinces, their approach achieved significant quality improvements while consistently satisfying operational constraints.

Several studies employed urban infrastructure as their testbed. Shafahi and Haghani [100] proposed a patrol routing model to ensure each road segment was covered multiple times per period, integrating workload balancing and location flexibility. They validated their model using the University of Maryland's road network. In a resilience-oriented study, [31] incorporated robustness against extreme events (e.g., terrorism, severe weather) into police deployment optimization, applying the model to the Greater London Area. Two studies by [75, 76] used real crime data from the Central District of Madrid to enhance patrol districting for predictive policing. The first reformulated the multi-covering patrol problem for generic graphs and proposed improved heuristics, while the latter introduced fairness-aware patrol planning, balancing racial and territorial equity.

Other studies leveraged real-world policing data to model patrol decision-making. Muaafa et al. [87] proposed a multi-objective patrol optimization framework, tested using crime data from Seattle's North Precinct between 2011 and 2014. Frohlich et al. [42] modeled patrols for security guards and cash-in-transit services, aiming to minimize costs while maximizing route variation. Their model, tested on 748 banks and ATMs in Vienna, Austria, found that using multiple shortest paths between POIs improved solution quality. Yang et al. [124] decomposed patrol operations into functional components–deterrence, resource allocation, and emergency response–and validated a coordination algorithm using precinct data from Beijing. Similarly, [114] introduced a city-scale patrol model to maximize police visibility while ensuring response time constraints. Their algorithm, based on problem decomposition and grafting, was validated with real patrol data from Foshan, China.

Finally, [67] advanced a smart policing framework for South Korea, integrating big data analytics and real-time traffic conditions to forecast crime risks and optimize patrol routes. Their model, benchmarked against conventional intuition-based methods, significantly reduced response times to emergencies by refining patrol coverage across high-risk urban grids.

3.2.2 Objective Functions

In this section, we examine the objective functions of models that employed deterministic optimization as their primary modeling framework. Based on the structure and number of objective functions, we classify these models into three main categories: single-objective, min-max objective, and multiple-objective formulations.

In single-objective optimization problems, optimality is assessed by directly comparing the values of the objective function across feasible solutions. These models typically aim to minimize or maximize a well-defined criterion, such as cost, travel time, or coverage. A notable subset of these models incorporates min–max structures, where the objective is to minimize the worst-case performance across a set of agents or routes. For instance, rather than minimizing total cost, the goal may be to minimize the maximum cost incurred by any single unit. In our study, we label such models as min–max single-objective formulations, distinguishing them from true multi-objective approaches that simultaneously optimize multiple conflicting criteria.

When multiple objective functions are present, the concept of dominance becomes essential in evaluating solution quality–particularly to distinguish non-dominated solutions that represent trade-offs where no objective can be improved without worsening another ([120]). Among the deterministic optimization models with multiple objectives in our review, we identified three major subcategories based on how trade-offs are handled:

- Scalarization-based formulations, such as the weighted sum method and goal programming, transform multiple objectives into a single objective by assigning importance weights or target levels to each criterion. These methods are intuitive and computationally tractable but require subjective input on weights or goals.
- Constraint-based formulations, including the ϵ-constraint method, approach the problem by optimizing one primary objective while treating the remaining objectives as constraints (bounded by threshold values). This technique is particularly useful for generating diverse Pareto-optimal solutions and exploring trade-off frontiers.
- Metaheuristic approaches, such as genetic algorithms, adaptive large neighborhood search (ALNS), and artificial immune systems, employ population-based or stochastic search mechanisms to approximate the Pareto front. These techniques are especially advantageous for large-scale or nonlinear problems where exact methods become computationally infeasible.

It is important to note that our classification of multiple objective functions is grounded in the context of the studies included in this review. For a broader and more in-depth discussion of multi-objective optimization in the literature, we refer readers to the comprehensive review by [101].

The majority of the models adopted a single objective function as summarized in Fig. 3.3, and there are several min-max optimization approaches. In [30], the authors modeled their MILP problem with the goal of minimizing the time required to cover the maximum area. They proposed a heuristic algorithm called the "time-priority immune clonal selection algorithm with optimization-revision" (TIAOR) to solve the problem. In [32], the bi-level objective function was defined as minimizing the maximum response time over basic networks. To solve this bi-level non-linear problem, they introduced an additional variable, transformed the problem into a mixed integer program, and used a branch-and-bound algorithm to solve it optimally. Maini et al. [84] proposed a MILP formulation to solve a problem consisting of heterogeneous robots in a continuous monitoring environment where the objective was to minimize the sum of the maximum cost of the robot path and the opening costs of the depots. In [116], a tabu search algorithm is proposed to guide the design of security estates. They formulated the problem using the min-max k-postmen framework, aiming to minimize the maximum route length. Their results demonstrate that the algorithm significantly improves upon previously applied routing policies.

Among the papers with multiple objective functions, eight introduced methods to combine multiple objectives into one. In [57], the authors developed a software package to

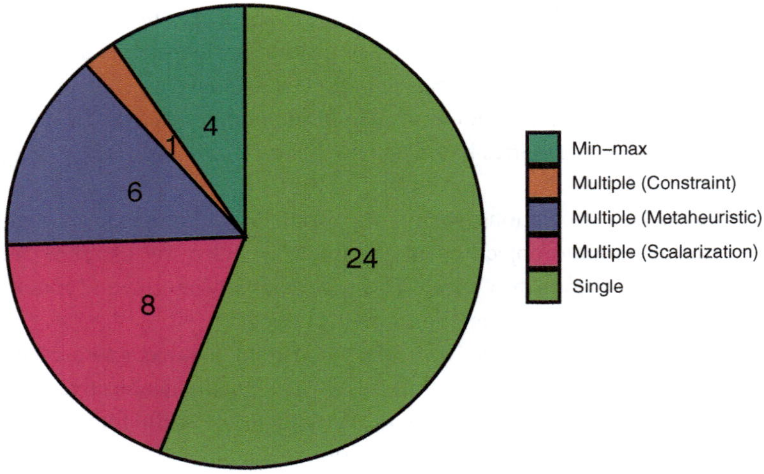

Fig. 3.3 Objective function types among papers that adopted deterministic optimization

optimize resource use across various scheduling missions. They introduced three optimization modules, with a total of eight different objective functions, and defined weighting factors to fine-tune the tradeoff between conflicting objectives. In [22], the authors initially defined three objectives, minimizing the number of bases, minimizing the patrolling resources, and maximizing the total coverage, which were then aggregated into a single composite objective function using a weighted sum approach and proposed an black-hole based algorithm to solve the model. Similarly, [62] defined two objectives: maximizing the coverage rate and minimizing task time. They introduced a utility function that incorporates both objectives, using a tuning parameter $\alpha \in [0, 1]$ to balance the tradeoff between them. They proposed a path planning algorithm WTSC, to solve the sweep coverage model and by comparing their algorithm's performance with benchmark algorithms CycleSplit ([44]) and G-MSCR ([81]), they reported an increase in coverage rate between 1 to 24 percent with their algorithm WTSC. Liberatore et al. [75] also utilized a scalar parameter: by controlling parameter $\lambda \in [0, 1]$, they aimed to balance patrolling optimization and workload balance within an objective function. They used different local search algorithms (Simple Hill Climbing algorithm, Steepest Descent Hill Climbing, and Tabu Search) to improve an initial solution, which is generated by a random-greedy algorithm. They reported a slight advance on produced results compared to previous benchmark models they cited. Qiao et al. [95] also introduced the parameters α and β, where $\alpha + \beta = 1$, to balance the objectives of total voyage distance and total finishing times of UAVs in the objective function. To solve it, they developed an algorithm that constructs patrol routes for UAVs by first computing the shortest paths using the Floyd algorithm and then incrementally inserting mandatory patrol nodes based on minimal insertion cost, ensuring that flight time and time window constraints are not violated. If a node cannot be inserted feasibly into an existing route, a new route is initialized, and the process repeats until all mandatory nodes are assigned. To balance the trade-off between maximizing total reward obtaining after successfully performing services and minimizing total working time, the authors in [112] also introduced α and β coefficients, formulating a single maximization problem where the latter (minimization) component is extracted from the former (maximization). They modeled the problem as an MILP and to solve large scale instances, they developed an iterated local search metaheuristic algorithm that starts with a solution built via a constructive heuristic, then repeatedly applies perturbations and Variable Neighborhood Descent-based local search, ensuring feasibility and quality of service constraints. When the objective function can be improved a new solution replaces the current one, until a time or iteration limit is reached. In [71], authors proposed 3 mathematical models to solve three different variants of the patrol scheduling problem. The second model included multiple objective functions in which the trade-off between two objective functions -maximizing the coverage and minimizing travel time- was controlled by a scalar parameter λ. Finally, in [76] authors proposed two objective functions: maximizing (i) the minimum proportion of service received across all districts (territorial fairness) and (ii) the minimum proportion of police coverage across all the population groups (racial fairness). Different from the previous research utilized scalarization technique with basically introducing scalar value(s), they utilized goal programming techniques to define ideal and anti-ideal values for each objective function, then proposed a weight parameter to illustrate decision maker's preference of the one objective over another by utilizing compromise programming techniques. They found that optimizing racial fairness aggressively resulted in very inefficient results; however even a tiny compromise of factor 0.01 in racial fairness improved the service level considerably.

The only paper utilized constraint-based methods to deal with multiple objective functions in our review is [2]. By extending their previous work [99], proposed a continuous drone surveillance model with two conflicting objectives: minimize the total length of E-line system and the number of assets. Since their model is not easy to model two conflicting objectives with a linear weighted objective function, or preemptive goal programming approaches, they formulated a multi-objective MINLP to generate Pareto-optimal solutions. In this approach, the objective of minimizing the total E-line length is treated as a constraint. To improve computational efficiency, they proposed an algorithm to identify critical drone speeds and a heuristic method to generate a high-quality initial solution before solving the exact model.

The remaining papers that used multiple objective functions in their models adopted metaheuristic approaches to deal with the trade-offs between conflicting objective functions. In [74], authors proposed a novel model to tackle with the disaster response system in a maritime environment. They modeled a multi-robot patrolling problem consisting of 3 different, conflicting objectives: minimize the total patrol length and patrolling time, as well as enforcing one of the patrolling robots to visit a target node that is within the range of the disaster location. To solve this multi-objective problem altogether, without solving them separately in a reasonable amount of time, they introduced a novel metaheuristic called Non-dominated Immune endocrine Short Feedback Algorithm (NISFA) to generate

Pareto-optimal solutions. Frohlich et al. [42] modeled a multi-objective problem with two conflicting objectives: minimizing cost and maximizing route inconsistency. For simplicity and ease of representation, they transformed the latter into minimizing route consistency. To address these conflicting objectives, they developed an ALNS-based metaheuristic to embed and analyze it into three different multi-objective frameworks (MDLS, ECH, EBSH). They reported that even a small shift favoring the cost-minimization objective resulted in significant deviations in consistency. Therefore, they suggested exercising caution when balancing these two conflicting objectives. Yanes Luis et al. [123] proposes a new graph-based approach to optimize multi-objective patrolling of water resources with autonomous surface vehicles, using tailored genetic operators for efficient solutions. To manage conflicting goals, they apply benchmark functions (Shekel, Himmelblau, and Rosenbrock) that mimic water quality parameters. They utilize Pareto front to help decision-makers balance vehicle count, travel distance, and patrolling performance. In [124], the authors modeled a multivariate relationship between patrol subjects, target points, and patrol deterrence. They defined three objectives: (i) minimizing assignment and patrolling costs for police vehicles and UAVs, (ii) minimizing the distribution differences in UAV tasks, and (iii) minimizing the distribution differences in police vehicle tasks and UAVs. This complex problem was analyzed using MILP formulations, genetic algorithms, and a new algorithm called the U-PBCPA is provided, which they used to evaluate the trade-offs between the objective functions. [86] and [87] also proposed pareto-optimal solutions for their multi-objective models and adopted a heuristic technique called PSDA to generate feasible solutions through Monte Carlo simulations. Using stochastic-dominance method, the conflicting objectives are sorted and ranked. At the end of MC iterations, the algorithm formulates the final approximation of Pareto set. [86] provided a synthetic case study in an imaginary border situation, whereas in [87] the authors designed a real-case study to generate best patrolling strategies for Seattle Police Department.

Other studies introduced single objective functions with following objectives: maximizing the expected number of successfully protected vehicles traveling [10], maximizing total patrol time at hot spots [21], maximizing the coverage [31, 100, 129], minimizing the time required to complete an operation [51, 118], minimizing total time not spent on patrol [33], minimizing the expected total response time [130], minimizing the travel time to arrive at hotspots [67] maximizing detection probability [34], minimizing total cost [27, 66, 81, 99], maximizing police visibility rate [114], [104], maximizing the total reward obtained [27, 41], minimizing complaint rate among customers [72] maximizing the number of potential threats inspected [29], minimize the average idleness of the graph [121], minimizing the distance between police stations [31] and minimizing the number of UAVs needed to patrol an area [56] (Table 3.4).

Table 3.4 Objective functions used in studies employing deterministic optimization as the primary modeling approach, organized by objective type: Single, Min-max, Multiple

Objective function	Objective function type	Papers
Minimize response/total travel time & Maximize hot-spot patrolling time	Single	[21, 51, 67, 70, 118, 130]
	Min-max	[30, 32]
	Multiple (Scalarization)	[62, 95, 112]
	Multiple (Metaheuristic)	[74]
Maximize coverage	Single	[31, 129]
	Min-max	[30]
	Multiple (Scalarization)	[22, 62, 71, 75]
Minimize costs	Single	[66, 81, 104]
	Min-max	[84]
	Multiple (Metaheuristic)	[42, 86, 87, 124]
Minimize distance	Single	[31]
	Min-max	[116]
	Multiple (Scalarization)	[71, 95]
	Multiple (Metaheuristic)	[74]
Maximize homogeneous workload distribution & homogeneity	Multiple (Scalarization)	[57, 75, 76]
	Multiple (Metaheuristic)	[124]
Minimize resources used	Single	[56]
	Multiple (Scalarization)	[22, 57]
	Multiple (ϵ-Constraint)	[2]
Maximize reward	Single	[27, 41]
	Multiple (Scalarization)	[112]
	Multiple (Metaheuristic)	[123]
Maximize satisfaction & minimize complaint rate	Single	[72]
	Multiple (Metaheuristic)	[87]
Maximize protected assets & assets inspected	Single	[10, 29]
Maximize police visibility rate	Single	[104, 114]
Minimize idleness	Single	[33, 121]
Minimize cost-benefit ratio	Multiple (Metaheuristic)	[87]
Maximize probability of detection	Single	[34]
Minimize vulnerability	Multiple (Metaheuristic)	[86]
Maximize route inconsistency	Multiple (Metaheuristic)	[42]

3.3 Analysis of Papers Adopted Stochastic Optimization as Their Primary Modeling Theme

In this section, we analyze the studies that primarily employ stochastic modeling approaches for patrolling problems. Table 3.5 provides a structured overview across six key dimensions: the domain of application (e.g., urban security, environmental monitoring), the type

Table 3.5 Overview of stochastic modeling approaches adopted in the reviewed studies, categorized by their domain, patrolling type, underlying modeling frameworks, sources of uncertainty, solution methods, and use of real-world case studies

Paper	Domain	Patrolling type	Source of uncertainty	Modeling framework	Solution method	Realistic case study
[107]	Theoretical	Regular	Infiltrator arrival time and location, dwell time	Stochastic Process (Poisson Process, Random Lifetimes)	Analytical solution	Not provided
[1]	Theoretical	Adversarial	Imperfect sensing capabilities	Markov Chain	Polynomial-time exact algorithm under min-max strategy	Not provided
[64]	Theoretical	Regular	Stochastic alert arrivals	MDP	Stochastic DP	Not provided
[68]	Maritime security	Adversarial	Detection success, target classification, transition and detection outcomes, interception timing and result	MDP	Stochastic DP, Backward DP algorithm	Provided
[106]	Theoretical	Regular	Noisy observations from regions; randomness in anomaly presence; stochastic routing	Stochastic routing & MC	Adaptive ensemble CUSUM algorithm	Not provided
[28]	Theoretical	Regular	Stochastic action selection using Boltzmann policy to introduce unpredictability in patrol routes	MDP	Cross-Entropy-based ϵ-optimal strategy	Not provided
[93]	Urban security	Regular	Robot failure rates, dynamic system state, distribution & cooperation of knowledge	Bayesian learning	AMCL & CBLS algorithms	Provided
[117]	Urban security	Regular	Imperfect sensoring, uncertainty in event locations	Decentralized POMDP	Image-based detection, data-fusion, MDP / POMDP planning	Provided
[14]	Theoretical	Adversarial	Adversarial behavior, partial observability of environment	POMDP	Evolutionary metaheuristic	Not provided
[77]	Theoretical	Adversarial	Randomized path selection to counter informed adversary	Markov chain	Tailored algorithms (FindPaths, PatrolRearrange)	Not provided
[119]	Theoretical	Adversarial	Partial observability; decentralized decisions	Decentralized POMDP	Decentralized Bayesian Reinforcement Learning (Monte Carlo Tree Search extension)	Not provided
[3]	Theoretic	Adversarial	Limited visibility, non-deterministic robot motion	Markov Chain	Convex optimization, tailored algorithm (CompleteVisibility)	Not provided
[82]	Environmental security	Regular	Environmental uncertainty, sensor noise, non-homogeneous reward function	MDP, DRL	Deep-Q-Learning, Double Deep Q-Learning algorithms	Provided
[129]	Theoretical	Regular	Environmental uncertainty, sensor noise, non-homogeneous reward function	POMDP	FB-SAMCP algorithm	Not provided

(continued)

Table 3.5 (continued)

Paper	Domain	Patrolling type	Source of uncertainty	Modeling framework	Solution method	Realistic case study
[98]	Urban security	Regular	Random requests, time-dependent travel times	Stochastic VRP	Recourse strategy-based simulation, MILP for a priori routing	Provided
[91]	Border security	Adversarial	Partially observable intruder locations	POMDP	Monte Carlo-based online planning	Provided
[122]	Maritime security	Regular	Partial observability of ocean environment, imperfect sensing capabilities	MDP, DRL	DDQN; LSTM-enhanced UANOA	Not provided
[115]	Urban security	Regular	Uncertain incident occurrence, uncertain travel time	MDP	OFFLINECSP (offline LP coordination), ONLINECSP (online heuristic policy)	Provided
[46]	Environmental security	Regular	Node visitation dynamics, UAV route adaptation in large areas, non-homogeneous reward function	MDP, DRL	DT-DRL algorithm	Not provided
[39]	Environmental security	Regular	Partial observability of pollution levels, environmental dynamics	POMDP	DDQ, DDQL, CDQL	Provided

of patrolling considered (regular or adversarial), the sources of uncertainty modeled (such as partial observability, stochastic alert arrivals, sensor noise, or adversary behavior), the underlying modeling framework used to represent this uncertainty (e.g., MDP, POMDP, Markov Chain, Bayesian learning), the solution methods adopted (including reinforcement learning, dynamic programming, Monte Carlo simulation, and problem-specific heuristics), and whether the study includes a realistic case study for validation.

Notably, most of the papers examined thus far either adopt Markov decision processes (MDP) or partially observable Markov decision processes (POMDP) as their core solution methodology or modeling framework. The only exceptions are [98], where the authors modeled their problem as a nonlinear stochastic integer program and used local search algorithms and the simulated annealing acceptance criterion to return the best solution encountered. Portugal and Rocha [93] developed their own Bayesian-inspired algorithm CBLS (Concurrent Bayesian Learning Strategy) where each robot's decision making progress built progressively based on the needs of the system for minimizing the objective function, minimizing average idleness. Finally, [107] derived analytical expressions for the detection rate based on two distinct patrol trajectories: leap-to-origin, which involves returning to a base location, and back-and-forth, which represents a cyclic movement between two endpoints. The remaining papers either adopted Markov chains [106], MDP [1, 3, 28, 64, 68, 82, 115, 122] or POMDP [14, 39, 91, 117, 119, 129] as their solution/modeling core.

Among the reviewed studies, seven papers adopted an adversarial patrolling perspective, explicitly modeling the interaction between patrollers and adversarial entities to optimize patrolling efforts. [1, 77], and [91] aimed to maximize the probability of detecting adversaries. While these studies acknowledged the importance of incorporating adversarial reasoning, they avoided traditional game-theoretic models in favor of tailored or heuristic-based solutions. Agmon et al. [1] proposed a min-max strategy to ensure minimum guaranteed detection, which was later extended in [77] to incorporate a reorganization phase under a sequential attack setting. In contrast, [91] adopted a POMDP approach in a realistic border surveillance context, modeling intruders as partially observable agents and assigning patrollers to zones. Notably, they validated their simulation with expert feedback, and their model outperformed baseline strategies in maximizing detection probability. Similarly, [3] addressed adversarial conditions by introducing non-deterministic patrols in limited-visibility environments. Their work minimized expected commute time through decentralized Markov chain policies and showed that individual randomized strategies could outperform centralized solutions. [14] also explored decentralized patrolling using POMDPs, modeling interactions between multiple patrollers and adversaries with the goal of maximizing illegal action detection. In a different vein, [68] presented a real-case interdiction scenario involving cooperative UAVs and ground interceptors. Their model addressed adversarial uncertainty in detection, classification, and interception success using stochastic dynamic programming. However, the complexity of the problem made it intractable, and they provided a simple greedy heuristic along with a relaxation of the model to offer lower and upper bounds for various problem sizes. The only adversarial patrolling paper that employed a reinforcement learning framework was [119], which modeled decentralized patrolling under unknown dynamics using a Bayes-adaptive POMDP. Each robot independently learned its patrolling policy via a Monte Carlo Tree Search extension and coordinated periodically to enhance information gathering. Their results showed superior performance compared to benchmark solvers in dynamic and uncertain environments.

By contrast, the regular patrolling studies did not include explicit adversarial behavior. In [107], the authors compared sensor placement strategies (back-and-forth vs. leap-to-origin) in a border surveillance context using stochastic process modeling to analyze trade-offs under varying cycle lengths. Several papers, including [28, 39, 46, 64, 82, 93, 122], and [115], adopted Markovian frameworks such as MDP, POMDP, or deep reinforcement learning to optimize regular patrol operations in uncertain environments. Among them, [28] modeled patrol planning as an MDP and used a cross-entropy algorithm to derive ϵ-optimal randomized patrol strategies for both finite and infinite-horizon cases. Kalyanam et al. [64] used stochastic dynamic programming to guide UAVs responding to alerts while balancing information gain and response time. Their method outperformed polling and greedy heuristics. In [93], the authors developed an online, decentralized framework to minimize node idleness in robot patrols, outperforming existing methods through a Bayesian learning-based algorithm.

In environmental and urban security domains, multiple papers explored deep reinforcement learning under stochastic conditions. Luis et al. [82] and [122] both used DQL and DDQL algorithms within MDP frameworks for patrolling water bodies, aiming to increase coverage and reduce revisit times. Similarly, [39] introduced a POMDP-based multitask learning approach to manage transitions between exploration and exploitation, outperforming both task-specific DQN models and heuristic baselines. Guan et al. [46] also addressed large-scale UAV routing via DRL, introducing trajectory diversity and dynamic-aware embeddings to improve patrol performance over complex terrain.

[115] contributed a hybrid two-stage framework for city-scale patrols. Their offline stage solved an LP using historical incident data to compute a benchmark patrol policy. The online component, modeled as an MDP, adapted to real-time requests using a competitive heuristic policy. Results confirmed its scalability and high incident response rates. Finally, [117] proposed an autonomous multi-objective surveillance system within a Decentralized POMDP framework. Their system balanced emergency response, visitor assistance, restricted-area security, and general patrolling, integrating perception and decision-making across agents.

References

1. Agmon, N., Kaminka, G. A., and Kraus, S. (2011). Multi-robot adversarial patrolling: Facing a full-knowledge opponent. *Journal of Artificial Intelligence Research*, 42:887–916.
2. Ahmadian, N., Lim, G. J., Torabbeigi, M., and Seon Jin, K. (2022). Smart border patrol using drones and wireless charging system under budget limitation. *Computers & Industrial Engineering*, 164:107891.
3. Alam, T., Rahman, M. M., Carrillo, P., Bobadilla, L., and Rapp, B. (2020). Stochastic multi-robot patrolling with limited visibility. *Journal of Intelligent & Robotic Systems*, 97(2):411–29.
4. Albarran, S. E. and Clempner, J. B. (2019). A stackelberg security markov game based on partial information for strategic decision making against unexpected attacks. *Engineering Applications of Artificial Intelligence*, 81:408–19.
5. Alpern, S., Bui, T., Lidbetter, T., and Papadaki, K. (2022). Continuous patrolling games. *Operations Research*, 70(6):3076–3089.
6. Alpern, S., Chleboun, P., Katsikas, S., and Lin, K. Y. (2021). Adversarial patrolling in a uniform. *Operations Research*, 70(1):129–140. https://doi.org/10.1287/opre.2021.2152.
7. Alpern, S., Lidbetter, T., and Papadaki, K. (2019). Optimizing periodic patrols against short attacks on the line and other networks. *European Journal of Operational Research*, 273(3):1065–73.
8. Alpern, S., Morton, A., and Papadaki, K. (2011). Patrolling games. *Operations Research*, 59(5):1246–1257. https://doi.org/10.1287/opre.2022.2346.
9. Atkinson, M. P., Kress, M., and Szechtman, R. (2017). To catch an intruder. part a: Uncluttered scenario. *Naval Research Logistics*, 64(1):29–40.
10. Auad, R. and Batta, R. (2017). Location-coverage models for preventing attacks on interurban transportation networks. *Annals of Operations Research*, 258(2):679–717.
11. Basilico, N., De Nittis, G., and Gatti, N. (2017). Adversarial patrolling with spatially uncertain alarm signals. *Artificial Intelligence*, 246:220–57.

12. Basilico, N., Gatti, N., and Amigoni, F. (2012). Patrolling security games: Definition and algorithms for solving large instances with single patroller and single intruder. *Artificial Intelligence*, 184-185:78–123.
13. Baykal-Gursoy, M., Duan, Z., Poor, H. V., and Garnaev, A. (2014). Infrastructure security games. *European Journal of Operational Research*, 239(2):469–478.
14. Beynier, A. (2017). A multiagent planning approach for cooperative patrolling with non-stationary adversaries. *International Journal on Artificial Intelligence Tools*, 26(5):28.
15. Brotcorne, L., Escalona, P., Fortz, B., and Labbé, M. (2021). Fare inspection patrols scheduling in transit systems using a stackelberg game approach. *Transportation Research Part B-Methodological*, 154:1–20.
16. Brown, G., Carlyle, M., Abdul-Ghaffar, A., and Kline, J. (2011). A defender-attacker optimization of port radar surveillance. *Naval Research Logistics*, 58(3):223–235.
17. Brown, G., Carlyle, M., Salmerón, J., and Wood, K. (2006). Defending critical infrastructure. *Interfaces*, 36(6):530–544.
18. Brown, M., Haskell, W. B., and Tambe, M. (2014). Addressing scalability and robustness in security games with multiple boundedly rational adversaries. *Lecture Notes in Computer Science (including subseries Lecture Notes in Artificial Intelligence and Lecture Notes in Bioinformatics)*, 8840:23–42.
19. Bucarey, L. V., Casorrán, C., Labbé, M., Ordoñez, F., and Figueroa, O. (2021). Coordinating resources in stackelberg security games. *European Journal of Operational Research*, 291(3):846–861.
20. Bui, T. and Lidbetter, T. (2023). Optimal patrolling strategies for trees and complete networks. *European Journal of Operational Research*, 311(2):769–776.
21. Capar, I., Keskin, B. B., and Rubin, P. A. (2015). An improved formulation for the maximum coverage patrol routing problem. *Computers & Operations Research*, 59:1–10.
22. Cardoso, F. A., Abrahao, F. T. M., and Saba, W. B. (2021). Resources package modelling supporting border surveillance operations. *Defence Science Journal*, 71(3):410–417.
23. Chao, L., Qian, Z., and Yanfeng, O. (2017). Planning of parking enforcement patrol considering drivers' parking payment behavior. *Transportation Research Part B: Methodological*, 106:375–92.
24. Chen, B., Zhu, Z. Q., Chen, F. R., Zhao, Y., and Qiu, X. G. (2019). Strategically patrolling in a chemical cluster addressing gas pollutants' releases through a game-theoretic model. *International Journal of Environmental Research and Public Health*, 16(4):18.
25. Chen, F., Chen, B., Zhu, Z., Wang, Y., Zhao, Y., Ji, Y., Qiu, X., Yin, Q., and Xiao, X. (2021a). An extended area-partition-involved collaborative patrolling game in chemical clusters considering attackers' bounded rationality and parameter uncertainty. *Process Safety and Environmental Protection*, 150:567–577.
26. Chen, F., Chen, B., Zhu, Z., Zhang, L., Qiu, X., Wang, Y., and Zhao, Y. (2021b). A cost-beneficial area-partition-involved collaborative patrolling game in a large-scale chemical cluster. *Process Safety and Environmental Protection*, 145:71–82.
27. Chen, X. (2012). Fast patrol route planning in dynamic environments. *IEEE Transactions on Systems, Man, and Cybernetics Part A:Systems and Humans*, 42(4):894–904.
28. Chen, X. (2013). Police patrol optimization with security level functions. *IEEE Transactions on Systems, Man, and Cybernetics Part A:Systems and Humans*, 43(5):1042–1051.
29. Chen, X. Y., Wu, S. N., Liu, Y. N., Wu, W. W., and Wang, S. A. (2022). A patrol routing problem for maritime crime-fighting. *Transportation Research Part E-Logistics and Transportation Review*, 168:19.
30. Cheng, L., Zhong, L., Tian, S., and Xing, J. (2019). Task assignment algorithm for road patrol by multiple uavs with multiple bases and rechargeable endurance. *IEEE Access*, 7:144381–144397.

31. Cheung, C. Y., Yoon, H. T., and Chow, A. H. F. (2015). Optimization of police facility deployment with a case study in greater london area. *Journal of Facilities Management*, 13(3):229–+.
32. Chih-Hsiung, H. and Egbelu, P. J. (2001). Selection of circulatory loops for patrol vehicles operating in a network. *International Journal of Production Research*, 39(4):579–601.
33. Chircop, P. A., Surendonk, T. J., van den Briel, M. H. L., and Walsh, T. (2022). On routing and scheduling a fleet of resource-constrained vessels to provide ongoing continuous patrol coverage. *Annals of Operations Research*, 312(2):723–760.
34. Chung, H., Polak, E., Royset, J. O., and Sastry, S. (2011). On the optimal detection of an underwater intruder in a channel using unmanned underwater vehicles. *Naval Research Logistics*, 58(8):804–820.
35. Clempner, J. B. (2018). A continuous-time markov stackelberg security game approach for reasoning about real patrol strategies. *International Journal of Control*, 91(11):2494–510.
36. Clempner, J. B. and Poznyak, A. S. (2020). A nucleus for bayesian partially observable markov games: joint observer and mechanism design. *Engineering Applications of Artificial Intelligence*, 95:381–92.
37. Delle Fave, F. M., Jiang, A. X., Yin, Z., Zhang, C., Tambe, M., Kraus, S., and Sullivan, J. P. (2014). Game-theoretic security patrolling with dynamic execution uncertainty and a case study on a real transit system. *Journal of Artificial Intelligence Research*, 50:321–367.
38. Delle Fave, F. M., Shieh, E., Jain, M., Jiang, A. X., Rosoff, H., Tambe, M., and Sullivan, J. P. (2015). Efficient solutions for joint activity based security games: fast algorithms, results and a field experiment on a transit system. *Autonomous Agents and Multi-Agent Systems*, 29(5):787–820.
39. Diop, D. S., Luis, S. Y., Esteve, M. P., Marin, S. L. T., and Reina, D. G. (2024). Decoupling patrolling tasks for water quality monitoring: A multi-agent deep reinforcement learning approach. *IEEE Access*, 12:75559–75576.
40. Escalona, P., Brotcorne, L., Fortz, B., and Ramirez, M. (2024). Fare inspection patrolling under in-station selective inspection policy. *Annals of Operations Research*, 332(1-3):191–212.
41. Farinelli, A., Iocchi, L., and Nardi, D. (2017). Distributed on-line dynamic task assignment for multi-robot patrolling. *Autonomous Robots*, 41(6):1321–45.
42. Frohlich, G. E. A., Doerner, K. F., and Gansterer, M. (2020). Secure and efficient routing on nodes, edges, and arcs of simple-graphs and of multi-graphs. *Networks*, 76(4):431–50.
43. Fu, H. and Liu, H. H. T. (2023). Defending a target area with a slower defender. *Ieee Control Systems Letters*, 7:661–666.
44. Gao, X., Fan, J., Wu, F., and Chen, G. (2018). Approximation algorithms for sweep coverage problem with multiple mobile sensors. *IEEE/ACM Transactions on Networking*, 26(2):990–1003.
45. Garrec, T. (2019). Continuous patrolling and hiding games. *European Journal of Operational Research*, 277(1):42–51.
46. Guan, Q., Zhang, X., Xie, M., Nie, J., Cao, H., Chen, Z., and He, Z. (2023). Large-scale power inspection: A deep reinforcement learning approach. *Frontiers in Energy Research*, 10.
47. Guerrero, D., Carsteanu, A. A., and Clempner, J. B. (2018). Solving stackelberg security markov games employing the bargaining nash approach: Convergence analysis. *Computers & Security*, 74:240–57.
48. Han, J. P., Wang, Z., Chen, X. G., Yang, M. Z., and Wang, F. Y. (2024). Digger: A graph contraction algorithm for patrolling games. *Ieee Transactions on Reliability*, 73(2):1104–1116.
49. Hazra, T., Kumar, C. R. S., and Nene, M. (2017). Multi-agent target searching with time constraints using game-theoretic approaches. *Kybernetes*, 46(8):1278–302.
50. Hazra, T., Kumar, C. R. S., and Nene, M. (2018). Two-agent cooperative search model with petri nets. *International Journal of Intelligent Unmanned Systems*, 6(4):162–173.

51. He, L., Peng, Z., Jiajie, W., Guoqiang, W., and Fanhe, M. (2019). Traffic patrolling routing problem with drones in an urban road system. *Sensors*, 19(23):5164 (20 pp.).
52. Hochbaum, D. S., Cheng, L., and Ordonez, F. (2014). Security routing games with multivehicle chinese postman problem. *Networks*, 64(3):181–91.
53. Hohzaki, R. (2011). Inspection game with smuggler's decision on the amount of contraband. *Journal of the Operations Research Society of Japan*, 54(1):25–45.
54. Hohzaki, R. and Maehara, H. (2010). A single-shot game of multi-period inspection. *European Journal of Operational Research*, 207(3):1410–18.
55. Hohzaki, R. and Masuda, R. (2012). A smuggling game with asymmetrical information of players. *Journal of the Operational Research Society*, 63(10):1434–1446.
56. Hoppmann-Baum, K., Burdakov, O., Mexi, G., Casselgren, C. J., and Koch, T. (2022). Length-constrained cycle partition with an application to uav routing*. *Optimization Methods and Software*, 37(6):2080–2116.
57. Horn, M. E. T., Jiang, H., and Kilby, P. (2007). Scheduling patrol boats and crews for the royal australian navy. *Journal of the Operational Research Society*, 58(10):1284–1293.
58. Huang, L., Zhou, M., Hao, K., and Hou, E. (2019). A survey of multi-robot regular and adversarial patrolling. *IEEE/CAA Journal of Automatica Sinica*, 6(4):894.
59. Hunt, K. and Zhuang, J. (2024). A review of attacker-defender games: Current state and paths forward. *European Journal of Operational Research*, 313(2):401–417.
60. Jain, M., Tsai, J., Pita, J., Kiekintveld, C., Rathi, S., Tambe, M., and Ordóñez, F. (2010). Software assistants for randomized patrol planning for the lax airport police and the federal air marshal service. *Interfaces*, 40(4):267–290.
61. Janssen, S., Matias, D., and Sharpanskykh, A. (2020). An agent-based empirical game theory approach for airport security patrols. *Aerospace*, 7(1):8 (22 pp.).
62. Jing, L., Yonghua, X., Jinhua, S., and Min, W. (2020). A path planning method for sweep coverage with multiple uavs. *IEEE Internet of Things Journal*, 7(9):8967–78.
63. Kahneman, D. and Tversky, A. (1979). Prospect theory: An analysis of decision under risk. *Econometrica*, 47(2):263–291.
64. Kalyanam, K., Chandler, P., Pachter, M., and Darbha, S. (2012). Optimization of perimeter patrol operations using unmanned aerial vehicles. *Journal of Guidance, Control, and Dynamics*, 35(2):434–41.
65. Karwowski, J. and Madziuk, J. (2019). A monte carlo tree search approach to finding efficient patrolling schemes on graphs. *European Journal of Operational Research*, 277(1):255–268.
66. Kaya, M. and Ozkan, O. (2021). Uav routing with genetic algorithm based matheuristic for border security missions. *International Journal of Optimization and Control-Theories & Applications-Ijocta*, 11(2):128–138.
67. Kim, D., Kan, Y., Aum, Y., Lee, W., and Yi, G. (2023). Hotspots-based patrol route optimization algorithm for smart policing. *Heliyon*, 9(10):16.
68. Kress, M., Royset, J. O., and Rozen, N. (2012). The eye and the fist: Optimizing search and interdiction. *European Journal of Operational Research*, 220(2):550–8.
69. Laan, C. M., Barros, A. I., Boucherie, R. J., Monsuur, H., and Timmer, J. (2019). Solving partially observable agent-intruder games with an application to border security problems. *Naval Research Logistics*, 66(2):174–190.
70. Lau, G. and Liu, H. H. T. (2014). Real-time path planning algorithm for autonomous border patrol: design, simulation, and experimentation. *Journal of Intelligent & Robotic Systems*, 75(3-4):517–39.
71. Lau, H. C., Yuan, Z., and Gunawan, A. (2016). Patrol scheduling in urban rail network. *Annals of Operations Research*, 239(1):317–342.
72. Lau, H. C. W., Ho, G. T. S., Yi, Z., and Hon, W. T. (2010). Optimizing patrol force deployment using a genetic algorithm. *Expert Systems with Applications*, 37(12):8148–54.

73. Lei, X., Hu, X., Wang, G., and Luo, H. (2023). A multi-uav deployment method for border patrolling based on stackelberg game. *Journal of Systems Engineering and Electronics*, pages 99–116.
74. Li, H., MengChu, Z., and Kuangrong, H. (2020). Non-dominated immune-endocrine short feedback algorithm for multi-robot maritime patrolling. *IEEE Transactions on Intelligent Transportation Systems*, 21(1):362–73.
75. Liberatore, F. and Camacho-Collados, M. (2016). A comparison of local search methods for the multicriteria police districting problem on graph. *Mathematical Problems in Engineering*, 2016.
76. Liberatore, F., Camacho-Collados, M., and Quijano-Sánchez, L. (2022). Equity in the police districting problem: Balancing territorial and racial fairness in patrolling operations. *Journal of Quantitative Criminology*, 38(3):697–721.
77. Lin, E. S., Agmon, N., and Kraus, S. (2019). Multi-robot adversarial patrolling: Handling sequential attacks. *Artificial Intelligence*, 274:1–25.
78. Lin, K. Y. (2022). Technical note–optimal patrol of a perimeter. *Operations Research*, 70(5):2860–2866.
79. Lin, K. Y., Atkinson, M. P., and Glazebrook, K. D. (2014). Optimal patrol to uncover threats in time when detection is imperfect. *Naval Research Logistics (NRL)*, 61(8):557–576.
80. Lin, K. Y., Atkinson, M. R., Chung, T. H., and Glazebrook, K. D. (2013). A graph patrol problem with random attack times. *Operations Research*, 61(3):694–710.
81. Liu, Y., Liu, Z., Shi, J., Wu, G., and Chen, C. (2019). Optimization of base location and patrol routes for unmanned aerial vehicles in border intelligence, surveillance, and reconnaissance. *Journal of Advanced Transportation*, 2019.
82. Luis, S. Y., Reina, D. G., and Marín, S. L. T. (2020). A deep reinforcement learning approach for the patrolling problem of water resources through autonomous surface vehicles: The ypacarai lake case. *Ieee Access*, 8:204076–204093.
83. Ma, W., McAreavey, K., Liu, W., and Luo, X. (2018). Acceptable costs of minimax regret equilibrium: A solution to security games with surveillance-driven probabilistic information. *Expert Systems with Applications*, 108:206–222.
84. Maini, P., Tokekar, P., and Sujit, P. B. (2021). Visibility-based persistent monitoring of piecewise linear features on a terrain using multiple aerial and ground robots. *IEEE Transactions on Automation Science and Engineering*, 18(4):1692–704.
85. McGrath, R. G. and Lin, K. Y. (2017). Robust patrol strategies against attacks at dispersed heterogeneous locations. *International Journal of Operational Research*, 30(3):340–59.
86. Muaafa, M. and Ramirez-Marquez, J. E. (2017a). Bi-objective evolutionary approach to the design of patrolling schemes for improved border security. *Computers & Industrial Engineering*, 107:74–84.
87. Muaafa, M. and Ramirez-Marquez, J. E. (2017b). Engineering management models for urban security. *IEEE Transactions on Engineering Management*, 64(1):29–41.
88. Oliva, G., Setola, R., and Tesei, M. (2019). A stackelberg game-theoretical approach to maritime counter-piracy. *IEEE Systems Journal*, 13(1):982–93.
89. Papadaki, K., Alpern, S., Lidbetter, T., and Morton, A. (2016). Patrolling a border. *Operations Research*, 64(6):1256–1269.
90. Paruchuri, P., Pearce, J. P., Marecki, J., Tambe, M., Ordonez, F., and Kraus, S. (2009). Coordinating randomized policies for increasing security of agent systems. *Information Technology & Management*, 10(1):67–79.
91. Pawgasame, W. and Wipusitwarakun, K. (2021). Mobile sensor patrol path planning in partially observable border regions. *Applied Intelligence*, 51(8):5453–5473.

92. Pita, J., Jain, M., Tambe, M., Ordóñez, F., and Kraus, S. (2010). Robust solutions to stackelberg games: Addressing bounded rationality and limited observations in human cognition. *Artificial Intelligence*, 174(15):1142–1171.
93. Portugal, D. and Rocha, R. P. (2016). Cooperative multi-robot patrol with bayesian learning. *Autonomous Robots*, 40(5):929–953.
94. Pulat, H. (2005). A two-sided optimization of border patrol interdiction. In *A Two-Sided Optimization of Border Patrol Interdiction*.
95. Qiao, Y., Yang, J., Zhang, Q., Xi, J., and Kong, L. (2019). Multi-uav cooperative patrol task planning novel method based on improved pfih algorithm. *IEEE Access*, 7:167621–167628.
96. Rezazadeh, A., Laobing, Z., Reniers, G., Khakzad, N., and Cozzani, V. (2017). Optimal patrol scheduling of hazardous pipelines using game theory. *Process Safety and Environmental Protection*, 109:242–56.
97. Royset, J. O. and Sato, H. (2010). Route optimization for multiple searchers. *Naval Research Logistics*, 57(8):701–717.
98. Saint-Guillain, M., Paquay, C., and Limbourg, S. (2021). Time-dependent stochastic vehicle routing problem with random requests: Application to online police patrol management in brussels. *European Journal of Operational Research*, 292(3):869–885.
99. Seon Jin, K. and Lim, G. J. (2018). Drone-aided border surveillance with an electrification line battery charging system. *Journal of Intelligent & Robotic Systems*, 92(3-4):657–70.
100. Shafahi, A. and Haghani, A. (2015). Generalized maximum benefit multiple chinese postman problem. *Transportation Research, Part C: Emerging Technologies*, 55:261–72.
101. Sharma, S. and Kumar, V. (2022). A comprehensive review on multi-objective optimization techniques: Past, present and future. *Archives of Computational Methods in Engineering*, 29(7):5605–5633.
102. Shen, W. R., Chen, W. Z., Huang, T. A., Singh, R., Xu, H. F., and Fang, F. (2024). An extensive study of security games with strategic informants. *Artificial Intelligence*, 334:22.
103. Simon, H. A. (1956). Rational choice and the structure of the environment. *Psychological review*, 63 2:129–38.
104. Snijders, H. and Saldanha, R. L. (2017). Decision support for scheduling security crews at netherlands railways. *Public Transport*, 9(1-2):193–215.
105. Solis, C. U., Clempner, J. B., and Poznyak, A. S. (2019). Handling a kullback-leibler divergence random walk for scheduling effective patrol strategies in stackelberg security games. *Kybernetika*, 55(4):618–640.
106. Srivastava, V., Pasqualetti, F., and Bullo, F. (2013). Stochastic surveillance strategies for spatial quickest detection. *International Journal of Robotics Research*, 32(12):1438–1458.
107. Szechtman, R., Kress, M., Kyle, L., and Cfir, D. (2008). Models of sensor operations for border surveillance. *Naval Research Logistics*, 55(1):27–41.
108. Tambe, M., Jain, M., and Tambe, M. (2011). *Introduction and Overview of Security Games*, page 1–24. Cambridge University Press.
109. Tokel, Y. I., Hunt, K., and Zhuang, J. (2025). Signaling the capability of new counterterrorism technologies in the face of a strategic threat. *Naval Research Logistics (NRL)*, 72(3):440–453.
110. Trejo, K. K., Clempner, J. B., and Poznyak, A. S. (2015). A stackelberg security game with random strategies based on the extraproximal theoretic approach. *Engineering Applications of Artificial Intelligence*, 37:145–53.
111. Trejo, K. K., Clempner, J. B., and Poznyak, A. S. (2018). Adapting attackers and defenders patrolling strategies: A reinforcement learning approach for stackelberg security games. *Journal of Computer and System Sciences*, 95:35–54.
112. Vidigal Correa, V. H., Dong, H., Iori, M., dos Santos, A. G., Yagiura, M., and Zucchi, G. (2024). An iterated local search for a multi-period orienteering problem arising in a car patrolling application. *Networks*, 83(1):153–168.

113. Wang, J. and Cui, L. (2023). Patrolling games with coordination between monitoring devices and patrols. *Reliability Engineering & System Safety*, page 109109.
114. Wang, W., Dong, Z., An, B., and Jiang, Y. (2021). Toward efficient city-scale patrol planning using decomposition and grafting. *IEEE Transactions on Intelligent Transportation Systems*, 22(2):747–757.
115. Wang, W., Tao, H., and Jiang, Y. (2022). Efficient online city-scale patrolling by exploiting offline model-based coordination policy. *IEEE Transactions on Intelligent Transportation Systems*, 23(8):13805–13818.
116. Willemse, E. J. and Joubert, J. W. (2012). Applying min-max k postmen problems to the routing of security guards. *Journal of the Operational Research Society*, 63(2):245–260.
117. Witwicki, S., Castillo, J. C., Messias, J., Capitan, J., Melo, F. S., Lima, P. U., and Veloso, M. (2017). Autonomous surveillance robots: A decision-making framework for networked muiltiagent systems. *IEEE Robotics and Automation Magazine*, 24(3):52–64.
118. Wu, G., Zhao, K., Cheng, J., and Ma, M. (2022). A coordinated vehicle–drone arc routing approach based on improved adaptive large neighborhood search. *Sensors*, 22(10).
119. Xin, Z., Weiping, W., Tao, W., Yonglin, L., and Fangcheng, Z. (2019). Bayesian reinforcement learning for multi-robot decentralized patrolling in uncertain environments. *IEEE Transactions on Vehicular Technology*, 68(12):11691–703.
120. Xiujuan, L. and Zhongke, S. (2004). Overview of multi-objective optimization methods. *Journal of Systems Engineering and Electronics*, 15(2):142–146.
121. Yan, C. B. and Zhang, T. (2016). Multi-robot patrol: A distributed algorithm based on expected idleness. *International Journal of Advanced Robotic Systems*, 13:12.
122. Yan, N., Huang, S., and Kong, C. (2021). Reinforcement learning-based autonomous navigation and obstacle avoidance for usvs under partially observable conditions. *Mathematical Problems in Engineering*, 2021.
123. Yanes Luis, S., Peralta, F., Tapia Cordoba, A., Rodriguez Del Nozal, A., Toral Marin, S., and Gutierrez Reina, D. (2022). An evolutionary multi-objective path planning of a fleet of asvs for patrolling water resources. *Engineering Applications of Artificial Intelligence*.
124. Yang, J., Zhaowei, D., and Lei, W. (2021). The programming model of air-ground cooperative patrol between multi-uav and police car. *IEEE Access*, 9:134503–134517.
125. Yang, R., Kiekintveld, C., Ordóñez, F., Tambe, M., and John, R. (2013). Improving resource allocation strategies against human adversaries in security games: An extended study. *Artificial Intelligence*, 195:440–469.
126. Yolme, A. and Baykal-Gürsoy, M. (2018). Urban rail patrolling: a game theoretic approach. *Journal of Transportation Security*, 11(1-2):23–40.
127. Yolmeh, A. and Baykal-Gürsoy, M. (2019). Patrolling games on general graphs with time dependent node values. *Military Operations Research*, 24(2):17–29.
128. Zhang, L. B., Reniers, G., Chen, B., and Qiu, X. G. (2019). Ccp game: A game theoretical model for improving the scheduling of chemical cluster patrolling. *Reliability Engineering & System Safety*, 191:12.
129. Zhou, X., Wang, W. P., Wang, T., Li, M., and Zhong, F. C. (2020). Online planning for multiagent situational information gathering in the markov environment. *Ieee Systems Journal*, 14(2):1798–1809.
130. Zhu, S. J., Kim, W., Chang, G. L., and Rochon, S. (2014). Design and evaluation of operational strategies for deploying emergency response teams: Dispatching or patrolling. *Journal of Transportation Engineering*, 140(6):12.
131. Zhuang, J. and Bier, V. M. (2011). Secrecy and deception at equilibrium, with applications to anti-terrorism resource allocation. *Defence and Peace Economics*, 22(1):43–61.

132. Zhuang, J., Bier, V. M., and Alagoz, O. (2010). Modeling secrecy and deception in a multiple-period attacker-defender signaling game. *European Journal of Operational Research*, 203(2):409–418.
133. Zoroa, N., Fernández-Sáez, M. J., and Zoroa, P. (2012). Patrolling a perimeter. *European Journal of Operational Research*, 222(3):571–582.

4 Future Research Directions

Out of the 122 papers we analyzed in this study, we identified 10 categories that authors repeatedly mentioned during their papers. We further observe that a portion of the reviewed studies did not offer any dedicated discussion, whether in a section, subsection, or within the conclusion, regarding possible future research directions arising from their limitations. Out of the remaining papers we analyzed, the following future research directions referenced the most, in order: Model expansion, relaxing modeling assumptions, algorithm/heuristic improvement, model validation, alternative model approaches, dynamic/online policies, vehicle/actor heterogeneity, data collection, multi-objective, and cooperation.

4.1 Model Expansion

A recurring theme in the Future Research Directions (FRD) sections was the intent to expand existing models by adding new features. The most frequently mentioned idea was incorporating uncertainty to improve adaptability. However, due to traceability and computational concerns, many studies maintained a deterministic framework (e.g., [14, 16, 21, 27, 49, 50, 52, 54, 58, 68, 71, 72, 76, 83, 89, 92]). Game-theoretical studies frequently proposed adding more players or units to reflect multi-agent dynamics, as seen in works like [7, 8, 19, 20, 34, 37, 54, 63], and [6].

Some researchers aimed to expand flexibility by turning parameters into decision variables or shifting from discrete to continuous variables. For example, [59, 63] and [14] proposed treating inspection time, crossing time and driver's designated parking lot parameters, respectively, as decision variables, while [5] suggested allowing attack locations to occur between nodes rather than at fixed points. Similarly, [90] proposed to refine their model

Y. I. Tokel and J. Zhuang, *Optimizing Security Patrolling Strategies*, Synthesis Lectures on Operations Research and Applications, https://doi.org/10.1007/978-3-032-02617-0_4

further by moving from binary protection (protected/not protected) measure to a continuous variable. Wang and Cui [80] also suggested considering the actual distances between targets, rather than classifying them simply as reachable or unreachable.

Others proposed adding realistic constraints to improve model robustness. For instance, [48] suggested incorporating qualitative factors; [18] considered introducing a response time limit; [85] proposed accounting for energy consumption and drone logistics; [23] explored the idea of including patrol boat maintenance cycles; and [94] discussed adding kinematic constraints to better reflect vehicle dynamics. Lin [53] and Oliva et al. [62] suggested incorporating additional information sources to enhance the defender's decision-making process. Additional ideas included modeling fatigue effects [48], generalizing to more network types [12], implementing the backup concept as a constraint rather than as an empty duty [75] and jointly considering task points and segments [21].

4.2 Relaxing Modeling Assumptions

Within game-theoretical frameworks, many authors have proposed relaxing built-in assumptions to reflect real-world behavior better. In our analysis, the two most frequently targeted assumptions are perfect rationality and complete information. Chen et al. [17], Pita et al. [66], Karwowski and Madziuk [43], Rezazadeh et al. [69], Wang and Cui [80], Janssen et al. [40] and Brotcorne et al. [10] all proposed to relax perfect rationality assumption stating that often human subjects do not act completely rationally and relaxing this critical assumption would mimic the human behavior much more realistically as stated in [74] and [66]. Similarly, recognizing that real players rarely have full knowledge of their opponents, several studies have recommended exploring imperfect or incomplete information variants. In addition to those already implemented (see Sect. 3.1.2), [14, 15, 38, 62, 73, 92], and [77] have all suggested extending their games in this direction.

Other than these two common assumptions, there are other assumptions that are targeted by authors. Instead of following a fixed temporal-spatial path, for example, [29] proposed to consider opportunistic passengers to mix these paths based on their domain knowledge, such as travel times and inspection probabilities. Zhang et al. [93] offered to relax the fixed patrolling time assumption to enable patrollers to spend varying times depending on the intensity of the location patrolled. Similarly, [4] and [87] proposed assigning different values to network nodes/vertices instead of treating them uniformly, and both [80] and [8] recommend moving beyond risk neutrality to capture varying attitudes originating from the difference in risk preferences. Instead of having the same severity for each call, [95] planned to include different severity of the calls to study its effect on the clearance time. Capar et al. [13] proposed to have a flexible number of units available within a timeframe. Srivastava et al. [76] offered to relax the assumption of having the anomalies in the same region all the time. Several authors have also suggested relaxing assumptions around sensing and communication. Papadaki et al. [63] proposed to study imperfect detection, while [67] suggested to relax perfect communication assumption.

Finally, to capture more complex scenarios, some studies proposed to relax the restricted area representation of their model. Hazra et al. [33, 34] and [44] proposed to relax the two-grid network space structure of their model to study three-dimensional spaces. Zhou et al. [94] suggested enlarging the grid size of their represented area to better mimic real-world applications.

4.3 Algorithm/Heuristic Improvement

The third most frequently observed future research direction was the improvement and development of algorithms and heuristics. Due to environmental complexity, most developed algorithms are effective only up to a certain instance size. To overcome scalability issues, many authors–particularly those with deterministic models–proposed either developing new algorithms or heuristics alongside their existing models or improving their current methods. The following studies suggested one or both approaches: [6, 7, 10, 15, 20–23, 30–32, 39, 46, 47, 51, 55, 56, 58, 67, 68, 82, 89, 93], and [86].

Developing approximation algorithms to better estimate solution bounds when exact optimization is computationally prohibitive was proposed by [6, 42, 58], and [39]. To handle very large instances, [21] argued for fast, feasible-solution generators, while [35] recommended methods to quickly prune the defender's strategy space. Meanwhile, [82] proposed an iterative policy-optimization scheme to reduce computational overhead. Zhang et al. [93] suggested developing algorithms or heuristics that yield robust solutions when optimal ones are infeasible due to environmental constraints. Chircop [23] and Lau et al. [47] recommended improving column-generation and subpricing procedures to handle larger instances more efficiently. Cheung et al. [22] argued for shifting from greedy methods to more advanced techniques–such as branch-and-bound or Lagrangian relaxation–to produce more reliable solutions. Finally, [51] stated they would seek algorithms that guarantee optimal solutions for their MC-PDP problem.

4.4 Model Validation

Although most studies validated their algorithms and solution techniques through simulation or numerical analysis, approximately one-fifth of the reviewed papers specifically stated the need for more tailored instances test and validate their approaches better. For instance, [26] proposed conducting a long-term controlled experiment to more deeply evaluate the strengths and limitations of their stochastic patrolling schedules. Similarly, some researchers highlighted the importance of validating their work in real-world scenarios rather than relying solely on synthetic or limited datasets. For example, [11] emphasized the need for more realistic case studies, while [36] pointed out the necessity of finding better representative cases to test specific aspects of their model. Likewise, [45] noted that access to recent local crime data might serve as a valuable validation resource to assess model performance under actual conditions.

A number of studies also proposed applying their models to different domains in order to evaluate generalizability and robustness. For instance, [25, 31, 40, 50, 64, 77, 84], and [7] recommended testing their models in new domains such as hospitals, schools, banks [40], the Internet of Things [50], shopping malls [84], shopping malls or bank branches [77], or across a broader range of smart grids [31]. These efforts were proposed to examine the adaptability of their models and collect operational feedback into performance variation across domains with different needs/constraints.

Human-subject experiments also emerged as a proposed direction for future validation efforts. While some studies like [66] and [90] already implemented such experiments to validate model assumptions and fine-tune parameters, others such as [9, 40], and [57] proposed adopting behavioral modeling approaches with human-subject data. These researchers argued that integrating empirical behavioral insights could enhance the credibility and applicability of their analytical or mathematical findings. Additionally, several works expressed the need to test models in real physical environments rather than controlled simulations. This was suggested by [24, 65, 78, 90], and [88], who emphasized the value of real-world implementation in understanding the practical challenges and operational viability of their models.

4.5 Alternative Modeling Approaches

When a particular modeling framework fails to capture a fundamental aspect of a problem, some researchers have proposed adopting entirely different modeling approaches to better reflect the complexities of the real world. Notably, many of these efforts also align with a broader aim of enhancing realism in problem representation. For instance, [61] proposed using agent-based modeling to better capture the dynamics of heterogeneous criminal behavior. Similarly, to more realistically represent adversarial profiles, [9] advocated transitioning from a purely stochastic approach to a behavioral game-theoretic framework. In an effort to emphasize coordination among team members operating in adversarial, non-cooperative settings, [57] suggested the use of cooperative games. Addressing uncertainty in the execution of patrolling tasks–such as delays–[26] recommended decentralized Markov decision processes. To capture the evolving knowledge state of players during interactions, [5] proposed modeling the problem as a differential or Markov game. Along similar lines, [15] and [93] proposed using stochastic games to address these dynamic elements. Additionally, studies such as [72, 81], and [79] focused on introducing stochastic elements to account for uncertainty in fixed parameters like travel times or the random arrival of incident requests.

Additionally, some researchers have emphasized the need to explicitly model the interactions among multiple decision-makers, rather than treating others' behaviors as fixed inputs. For example, [3] proposed adopting a game-theoretic framework to explore the adversary's decision-making process more thoroughly. Similarly, [51] noted the potential value of game theory in analyzing reciprocal strategies among interacting agents. To better study strategic adversarial behavior, [47] proposed the use of the Stackelberg security game framework. Additionally, [91] suggested a shift from a simultaneous move structure to a sequential one, wit the objective of more accurately capturing the implications of transparency and timing in adversarial interactions.

4.6 Dynamic/Online Policies

To extend their static models, several authors have proposed developing online (or dynamic) policies. Fundamentally, static and online approaches differ: static policies require a complete data set up front in order to solve the problem optimally, whereas dynamic policies periodically update incoming data and resolve the problem–using techniques such as dynamic programming, reinforcement learning, or adaptive heuristics.

Because dynamic environments are far more common in real–world applications, these authors underlined the need for deeper study of online and dynamic policies. For example, [50] were among the first to transplant their NISFA framework into a fully dynamic setting, and [2] proposed to develop a location-based, time-varying risk model to re-optimize drone patrols as new threat information arrives. Similarly, [85] argued for modeling demand as a stochastic, evolving process to build greater robustness against disruptions.

A related strand focuses specifically on real-time path planning and patrol scheduling. Jing et al. [41], Vidigal Correa et al. [79], and Wang et al. [81] each demonstrated that incorporating live data, including but not limited to unexpected alarms, urgent call-outs, or up-to-the-minute traffic flows into recursive model updates significantly improve adaptability compared to fixed offline routes. Brotcorne et al. [10] further generalized this idea to dynamically schedule inspections, while [35] and [55] pointed out to incorporate dynamic time windows and environment-dependent detection rules as potential future research ideas, respectively.

Finally, the authors who developed game-theoretic formulations highlighted the importance of online interaction. Both [90] and [54] proposed that attackers observe evolving conditions and potential targets before acting, and [70] later assessed whether the gains from such interactivity outweigh the additional computational complexity. As seen clearly, these "future research" ideas and propositions underscore a common consensus: to handle the complexities of real-world defense and security operations, models must learn, adapt, and re-solve as fresh data become available.

4.7 Vehicle/Actor Heterogeneity

To obtain more tractable solutions, authors often make simplifying assumptions that reduce model dimensionality. For example, in game theory, the assumption of complete information is commonly made because incorporating incomplete information typically leads to more complex models, making analytical solutions more difficult. Similarly, when modeling multiple units, their characteristics are often assumed to be homogeneous–such as a fleet of drones with identical speed and range–despite potential differences among units of the same type.

Several authors underlined the need to incorporate actor or vehicle heterogeneity in future research. Agmon et al. [1] and Farinelli et al. [30] emphasized the importance of handling heterogeneous robots. Srivastava et al. [76] suggested studying time-varying Markov chains, noting that they might offer faster convergence to the desired stationary distribution compared to the adopted time-homogeneous models. McGrath and Lin [59] recognized the potential differences between patrollers and proposed studying heterogeneous cases to better capture these variations. Muaafa and Ramirez-Marquez [61] acknowledged the limitation of modeling only homogeneous criminals and suggested exploring agent-based approaches to more effectively represent environments with heterogeneous agents. Yanes Luis et al. [89] recommended investigating heterogeneous multi-vehicle systems to dynamically allocate tasks based on vehicle-specific advantages. Similarly, [23] proposed extending their model by considering a heterogeneous fleet of patrol boats, rather than assuming a single vessel class with identical characteristics.

4.8 Data Collection

As expected, several authors expressed concerns about the lack of available data needed to better validate and refine their approaches. Lau and Liu [45] emphasized the need for comprehensive, accumulated data on actual crime incidents in order to analyze the correlation between crime and various dependent variables related to local community characteristics. Similarly, [65] highlighted the importance of richer data sources to enable more efficient data processing. In addition to these points, [60, 84], and [88] shed light on another critical role of data: improving the estimation of key model parameters. They proposed gathering additional information to better estimate state features, transition probabilities, and reward structures [84], sensor-based inputs such as images [88], and probabilities related to illicit crossings [60].

4.9 Multi-objective

When selecting an objective function to represent their problem, researchers often choose among several closely competing candidates. Although Sect. 3.2.1 highlighted that many papers in this survey already propose multiple objective functions to define their models, several studies identified the need to consider multiple objectives to better capture the complexities involved. Capar et al. [13] suggested including additional duties for troopers beyond hot spot coverage. Diop et al. [28] expressed interest in moving away from predefined task durations toward training the network with multiple objectives. Wu et al. [85] proposed exploring the trade-off between conflicting goals, such as minimizing total costs and maximizing patrol efficiency. Similarly, [91] highlighted the importance of studying the trade-offs between operational and strategic decisions through a hybrid approach, noting that this could significantly reduce expected losses.

4.10 Cooperation

Although cooperation between patrolling resources is crucial for optimal patrolling operations, only three papers in our analysis of non-cooperative modeling approaches explicitly identified coordination as a future research direction. Lin [53] highlighted the importance of scenarios where an attacker collaborates with an accomplice, and how defenders might respond to such cooperation. Chen [19] suggested studying cooperation between patrollers within a single patrol beat. Finally, [30] proposed extending their online coordination algorithms to account for scenarios in which strategic reasoning influences cumulative decision-making and encourages joint actions among robots.

References

1. Agmon, N., Kaminka, G. A., and Kraus, S. (2011). Multi-robot adversarial patrolling: Facing a full-knowledge opponent. *Journal of Artificial Intelligence Research*, 42:887–916.
2. Ahmadian, N., Lim, G. J., Torabbeigi, M., and Seon Jin, K. (2022). Smart border patrol using drones and wireless charging system under budget limitation. *Computers & Industrial Engineering*, 164:107891.
3. Alam, T., Rahman, M. M., Carrillo, P., Bobadilla, L., and Rapp, B. (2020). Stochastic multi-robot patrolling with limited visibility. *Journal of Intelligent & Robotic Systems*, 97(2):411–29.
4. Alpern, S., Bui, T., Lidbetter, T., and Papadaki, K. (2022). Continuous patrolling games. *Operations Research*, 70(6):3076–3089.
5. Alpern, S., Morton, A., and Papadaki, K. (2011). Patrolling games. *Operations Research*, 59(5):1246–1257. https://doi.org/10.1287/opre.2022.2346.
6. Basilico, N., De Nittis, G., and Gatti, N. (2017). Adversarial patrolling with spatially uncertain alarm signals. *Artificial Intelligence*, 246:220–57.
7. Basilico, N., Gatti, N., and Amigoni, F. (2012). Patrolling security games: Definition and algorithms for solving large instances with single patroller and single intruder. *Artificial Intelligence*, 184-185:78–123.
8. Baykal-Gursoy, M., Duan, Z., Poor, H. V., and Garnaev, A. (2014). Infrastructure security games. *European Journal of Operational Research*, 239(2):469–478.
9. Beynier, A. (2017). A multiagent planning approach for cooperative patrolling with non-stationary adversaries. *International Journal on Artificial Intelligence Tools*, 26(5):28.

10. Brotcorne, L., Escalona, P., Fortz, B., and Labbé, M. (2021). Fare inspection patrols scheduling in transit systems using a stackelberg game approach. *Transportation Research Part B-Methodological*, 154:1–20.
11. Bucarey, L. V., Casorrán, C., Labbé, M., Ordoñez, F., and Figueroa, O. (2021). Coordinating resources in stackelberg security games. *European Journal of Operational Research*, 291(3):846–861.
12. Bui, T. and Lidbetter, T. (2023). Optimal patrolling strategies for trees and complete networks. *European Journal of Operational Research*, 311(2):769–776.
13. Capar, I., Keskin, B. B., and Rubin, P. A. (2015). An improved formulation for the maximum coverage patrol routing problem. *Computers & Operations Research*, 59:1–10.
14. Chao, L., Qian, Z., and Yanfeng, O. (2017). Planning of parking enforcement patrol considering drivers' parking payment behavior. *Transportation Research Part B: Methodological*, 106:375–92.
15. Chen, B., Zhu, Z. Q., Chen, F. R., Zhao, Y., and Qiu, X. G. (2019). Strategically patrolling in a chemical cluster addressing gas pollutants' releases through a game-theoretic model. *International Journal of Environmental Research and Public Health*, 16(4):18.
16. Chen, F., Chen, B., Zhu, Z., Wang, Y., Zhao, Y., Ji, Y., Qiu, X., Yin, Q., and Xiao, X. (2021a). An extended area-partition-involved collaborative patrolling game in chemical clusters considering attackers' bounded rationality and parameter uncertainty. *Process Safety and Environmental Protection*, 150:567–577.
17. Chen, F., Chen, B., Zhu, Z., Zhang, L., Qiu, X., Wang, Y., and Zhao, Y. (2021b). A cost-beneficial area-partition-involved collaborative patrolling game in a large-scale chemical cluster. *Process Safety and Environmental Protection*, 145:71–82.
18. Chen, X. (2012). Fast patrol route planning in dynamic environments. *IEEE Transactions on Systems, Man, and Cybernetics Part A:Systems and Humans*, 42(4):894–904.
19. Chen, X. (2013). Police patrol optimization with security level functions. *IEEE Transactions on Systems, Man, and Cybernetics Part A:Systems and Humans*, 43(5):1042–1051.
20. Chen, X. Y., Wu, S. N., Liu, Y. N., Wu, W. W., and Wang, S. A. (2022). A patrol routing problem for maritime crime-fighting. *Transportation Research Part E-Logistics and Transportation Review*, 168:19.
21. Cheng, L., Zhong, L., Tian, S., and Xing, J. (2019). Task assignment algorithm for road patrol by multiple uavs with multiple bases and rechargeable endurance. *IEEE Access*, 7:144381–144397.
22. Cheung, C. Y., Yoon, H. T., and Chow, A. H. F. (2015). Optimization of police facility deployment with a case study in greater london area. *Journal of Facilities Management*, 13(3):229–+.
23. Chircop, P. A., Surendonk, T. J., van den Briel, M. H. L., and Walsh, T. (2022). On routing and scheduling a fleet of resource-constrained vessels to provide ongoing continuous patrol coverage. *Annals of Operations Research*, 312(2):723–760.
24. Clempner, J. B. (2018). A continuous-time markov stackelberg security game approach for reasoning about real patrol strategies. *International Journal of Control*, 91(11):2494–510.
25. Clempner, J. B. and Poznyak, A. S. (2020). A nucleus for bayesian partially observable markov games: joint observer and mechanism design. *Engineering Applications of Artificial Intelligence*, 95:381–92.
26. Delle Fave, F. M., Jiang, A. X., Yin, Z., Zhang, C., Tambe, M., Kraus, S., and Sullivan, J. P. (2014). Game-theoretic security patrolling with dynamic execution uncertainty and a case study on a real transit system. *Journal of Artificial Intelligence Research*, 50:321–367.
27. Delle Fave, F. M., Shieh, E., Jain, M., Jiang, A. X., Rosoff, H., Tambe, M., and Sullivan, J. P. (2015). Efficient solutions for joint activity based security games: fast algorithms, results and a field experiment on a transit system. *Autonomous Agents and Multi-Agent Systems*, 29(5):787–820.

28. Diop, D. S., Luis, S. Y., Esteve, M. P., Marin, S. L. T., and Reina, D. G. (2024). Decoupling patrolling tasks for water quality monitoring: A multi-agent deep reinforcement learning approach. *IEEE Access*, 12:75559–75576.
29. Escalona, P., Brotcorne, L., Fortz, B., and Ramirez, M. (2024). Fare inspection patrolling under in-station selective inspection policy. *Annals of Operations Research*, 332(1-3):191–212.
30. Farinelli, A., Iocchi, L., and Nardi, D. (2017). Distributed on-line dynamic task assignment for multi-robot patrolling. *Autonomous Robots*, 41(6):1321–45.
31. Guan, Q., Zhang, X., Xie, M., Nie, J., Cao, H., Chen, Z., and He, Z. (2023). Large-scale power inspection: A deep reinforcement learning approach. *Frontiers in Energy Research*, 10.
32. Han, J. P., Wang, Z., Chen, X. G., Yang, M. Z., and Wang, F. Y. (2024). Digger: A graph contraction algorithm for patrolling games. *Ieee Transactions on Reliability*, 73(2):1104–1116.
33. Hazra, T., Kumar, C. R. S., and Nene, M. (2017). Multi-agent target searching with time constraints using game-theoretic approaches. *Kybernetes*, 46(8):1278–302.
34. Hazra, T., Kumar, C. R. S., and Nene, M. (2018). Two-agent cooperative search model with petri nets. *International Journal of Intelligent Unmanned Systems*, 6(4):162–173.
35. He, L., Peng, Z., Jiajie, W., Guoqiang, W., and Fanhe, M. (2019). Traffic patrolling routing problem with drones in an urban road system. *Sensors*, 19(23):5164 (20 pp.).
36. Hochbaum, D. S., Cheng, L., and Ordonez, F. (2014). Security routing games with multivehicle chinese postman problem. *Networks*, 64(3):181–91.
37. Hohzaki, R. (2011). Inspection game with smuggler's decision on the amount of contraband. *Journal of the Operations Research Society of Japan*, 54(1):25–45.
38. Hohzaki, R. and Maehara, H. (2010). A single-shot game of multi-period inspection. *European Journal of Operational Research*, 207(3):1410–18.
39. Hoppmann-Baum, K., Burdakov, O., Mexi, G., Casselgren, C. J., and Koch, T. (2022). Length-constrained cycle partition with an application to uav routing*. *Optimization Methods and Software*, 37(6):2080–2116.
40. Janssen, S., Matias, D., and Sharpanskykh, A. (2020). An agent-based empirical game theory approach for airport security patrols. *Aerospace*, 7(1):8 (22 pp.).
41. Jing, L., Yonghua, X., Jinhua, S., and Min, W. (2020). A path planning method for sweep coverage with multiple uavs. *IEEE Internet of Things Journal*, 7(9):8967–78.
42. Kalyanam, K., Chandler, P., Pachter, M., and Darbha, S. (2012). Optimization of perimeter patrol operations using unmanned aerial vehicles. *Journal of Guidance, Control, and Dynamics*, 35(2):434–41.
43. Karwowski, J. and Madziuk, J. (2019). A monte carlo tree search approach to finding efficient patrolling schemes on graphs. *European Journal of Operational Research*, 277(1):255–268.
44. Kaya, M. and Ozkan, O. (2021). Uav routing with genetic algorithm based matheuristic for border security missions. *International Journal of Optimization and Control-Theories & Applications-Ijocta*, 11(2):128–138.
45. Kim, D., Kan, Y., Aum, Y., Lee, W., and Yi, G. (2023). Hotspots-based patrol route optimization algorithm for smart policing. *Heliyon*, 9(10):16.
46. Lau, G. and Liu, H. H. T. (2014). Real-time path planning algorithm for autonomous border patrol: design, simulation, and experimentation. *Journal of Intelligent & Robotic Systems*, 75(3-4):517–39.
47. Lau, H. C., Yuan, Z., and Gunawan, A. (2016). Patrol scheduling in urban rail network. *Annals of Operations Research*, 239(1):317–342.
48. Lau, H. C. W., Ho, G. T. S., Yi, Z., and Hon, W. T. (2010). Optimizing patrol force deployment using a genetic algorithm. *Expert Systems with Applications*, 37(12):8148–54.
49. Lei, X., Hu, X., Wang, G., and Luo, H. (2023). A multi-uav deployment method for border patrolling based on stackelberg game. *Journal of Systems Engineering and Electronics*, pages 99–116.

50. Li, H., MengChu, Z., and Kuangrong, H. (2020). Non-dominated immune-endocrine short feedback algorithm for multi-robot maritime patrolling. *IEEE Transactions on Intelligent Transportation Systems*, 21(1):362–73.
51. Liberatore, F. and Camacho-Collados, M. (2016). A comparison of local search methods for the multicriteria police districting problem on graph. *Mathematical Problems in Engineering*, 2016.
52. Lin, E. S., Agmon, N., and Kraus, S. (2019). Multi-robot adversarial patrolling: Handling sequential attacks. *Artificial Intelligence*, 274:1–25.
53. Lin, K. Y. (2022). Technical note–optimal patrol of a perimeter. *Operations Research*, 70(5):2860–2866.
54. Lin, K. Y., Atkinson, M. R., Chung, T. H., and Glazebrook, K. D. (2013). A graph patrol problem with random attack times. *Operations Research*, 61(3):694–710.
55. Liu, Y., Liu, Z., Shi, J., Wu, G., and Chen, C. (2019). Optimization of base location and patrol routes for unmanned aerial vehicles in border intelligence, surveillance, and reconnaissance. *Journal of Advanced Transportation*, 2019.
56. Luis, S. Y., Reina, D. G., and Marín, S. L. T. (2020). A deep reinforcement learning approach for the patrolling problem of water resources through autonomous surface vehicles: The ypacarai lake case. *Ieee Access*, 8:204076–204093.
57. Ma, W., McAreavey, K., Liu, W., and Luo, X. (2018). Acceptable costs of minimax regret equilibrium: A solution to security games with surveillance-driven probabilistic information. *Expert Systems with Applications*, 108:206–222.
58. Maini, P., Tokekar, P., and Sujit, P. B. (2021). Visibility-based persistent monitoring of piecewise linear features on a terrain using multiple aerial and ground robots. *IEEE Transactions on Automation Science and Engineering*, 18(4):1692–704.
59. McGrath, R. G. and Lin, K. Y. (2017). Robust patrol strategies against attacks at dispersed heterogeneous locations. *International Journal of Operational Research*, 30(3):340–59.
60. Muaafa, M. and Ramirez-Marquez, J. E. (2017a). Bi-objective evolutionary approach to the design of patrolling schemes for improved border security. *Computers & Industrial Engineering*, 107:74–84.
61. Muaafa, M. and Ramirez-Marquez, J. E. (2017b). Engineering management models for urban security. *IEEE Transactions on Engineering Management*, 64(1):29–41.
62. Oliva, G., Setola, R., and Tesei, M. (2019). A stackelberg game-theoretical approach to maritime counter-piracy. *IEEE Systems Journal*, 13(1):982–93.
63. Papadaki, K., Alpern, S., Lidbetter, T., and Morton, A. (2016). Patrolling a border. *Operations Research*, 64(6):1256–1269.
64. Paruchuri, P., Pearce, J. P., Marecki, J., Tambe, M., Ordonez, F., and Kraus, S. (2009). Coordinating randomized policies for increasing security of agent systems. *Information Technology & Management*, 10(1):67–79.
65. Pawgasame, W. and Wipusitwarakun, K. (2021). Mobile sensor patrol path planning in partially observable border regions. *Applied Intelligence*, 51(8):5453–5473.
66. Pita, J., Jain, M., Tambe, M., Ordóñez, F., and Kraus, S. (2010). Robust solutions to stackelberg games: Addressing bounded rationality and limited observations in human cognition. *Artificial Intelligence*, 174(15):1142–1171.
67. Portugal, D. and Rocha, R. P. (2016). Cooperative multi-robot patrol with bayesian learning. *Autonomous Robots*, 40(5):929–953.
68. Qiao, Y., Yang, J., Zhang, Q., Xi, J., and Kong, L. (2019). Multi-uav cooperative patrol task planning novel method based on improved pfih algorithm. *IEEE Access*, 7:167621–167628.
69. Rezazadeh, A., Laobing, Z., Reniers, G., Khakzad, N., and Cozzani, V. (2017). Optimal patrol scheduling of hazardous pipelines using game theory. *Process Safety and Environmental Protection*, 109:242–56.

70. Saint-Guillain, M., Paquay, C., and Limbourg, S. (2021). Time-dependent stochastic vehicle routing problem with random requests: Application to online police patrol management in brussels. *European Journal of Operational Research*, 292(3):869–885.
71. Seon Jin, K. and Lim, G. J. (2018). Drone-aided border surveillance with an electrification line battery charging system. *Journal of Intelligent & Robotic Systems*, 92(3-4):657–70.
72. Shafahi, A. and Haghani, A. (2015). Generalized maximum benefit multiple chinese postman problem. *Transportation Research, Part C: Emerging Technologies*, 55:261–72.
73. Shen, W. R., Chen, W. Z., Huang, T. A., Singh, R., Xu, H. F., and Fang, F. (2024). An extensive study of security games with strategic informants. *Artificial Intelligence*, 334:22.
74. Simon, H. A. (1956). Rational choice and the structure of the environment. *Psychological review*, 63 2:129–38.
75. Snijders, H. and Saldanha, R. L. (2017). Decision support for scheduling security crews at netherlands railways. *Public Transport*, 9(1-2):193–215.
76. Srivastava, V., Pasqualetti, F., and Bullo, F. (2013). Stochastic surveillance strategies for spatial quickest detection. *International Journal of Robotics Research*, 32(12):1438–1458.
77. Trejo, K. K., Clempner, J. B., and Poznyak, A. S. (2015). A stackelberg security game with random strategies based on the extraproximal theoretic approach. *Engineering Applications of Artificial Intelligence*, 37:145–53.
78. Trejo, K. K., Clempner, J. B., and Poznyak, A. S. (2018). Adapting attackers and defenders patrolling strategies: A reinforcement learning approach for stackelberg security games. *Journal of Computer and System Sciences*, 95:35–54.
79. Vidigal Correa, V. H., Dong, H., Iori, M., dos Santos, A. G., Yagiura, M., and Zucchi, G. (2024). An iterated local search for a multi-period orienteering problem arising in a car patrolling application. *Networks*, 83(1):153–168.
80. Wang, J. and Cui, L. (2023). Patrolling games with coordination between monitoring devices and patrols. *Reliability Engineering & System Safety*, page 109109.
81. Wang, W., Dong, Z., An, B., and Jiang, Y. (2021). Toward efficient city-scale patrol planning using decomposition and grafting. *IEEE Transactions on Intelligent Transportation Systems*, 22(2):747–757.
82. Wang, W., Tao, H., and Jiang, Y. (2022). Efficient online city-scale patrolling by exploiting offline model-based coordination policy. *IEEE Transactions on Intelligent Transportation Systems*, 23(8):13805–13818.
83. Willemse, E. J. and Joubert, J. W. (2012). Applying min-max k postmen problems to the routing of security guards. *Journal of the Operational Research Society*, 63(2):245–260.
84. Witwicki, S., Castillo, J. C., Messias, J., Capitan, J., Melo, F. S., Lima, P. U., and Veloso, M. (2017). Autonomous surveillance robots: A decision-making framework for networked muiltiagent systems. *IEEE Robotics and Automation Magazine*, 24(3):52–64.
85. Wu, G., Zhao, K., Cheng, J., and Ma, M. (2022). A coordinated vehicle–drone arc routing approach based on improved adaptive large neighborhood search. *Sensors*, 22(10).
86. Xin, Z., Weiping, W., Tao, W., Yonglin, L., and Fangcheng, Z. (2019). Bayesian reinforcement learning for multi-robot decentralized patrolling in uncertain environments. *IEEE Transactions on Vehicular Technology*, 68(12):11691–703.
87. Yan, C. B. and Zhang, T. (2016). Multi-robot patrol: A distributed algorithm based on expected idleness. *International Journal of Advanced Robotic Systems*, 13:12.
88. Yan, N., Huang, S., and Kong, C. (2021). Reinforcement learning-based autonomous navigation and obstacle avoidance for usvs under partially observable conditions. *Mathematical Problems in Engineering*, 2021.
89. Yanes Luis, S., Peralta, F., Tapia Cordoba, A., Rodriguez Del Nozal, A., Toral Marin, S., and Gutierrez Reina, D. (2022). An evolutionary multi-objective path planning of a fleet of asvs for patrolling water resources. *Engineering Applications of Artificial Intelligence*.

90. Yang, R., Kiekintveld, C., Ordóñez, F., Tambe, M., and John, R. (2013). Improving resource allocation strategies against human adversaries in security games: An extended study. *Artificial Intelligence*, 195:440–469.
91. Yolme, A. and Baykal-Gürsoy, M. (2018). Urban rail patrolling: a game theoretic approach. *Journal of Transportation Security*, 11(1-2):23–40.
92. Yolmeh, A. and Baykal-Gürsoy, M. (2019). Patrolling games on general graphs with time dependent node values. *Military Operations Research*, 24(2):17–29.
93. Zhang, L. B., Reniers, G., Chen, B., and Qiu, X. G. (2019). Ccp game: A game theoretical model for improving the scheduling of chemical cluster patrolling. *Reliability Engineering & System Safety*, 191:12.
94. Zhou, X., Wang, W. P., Wang, T., Li, M., and Zhong, F. C. (2020). Online planning for multiagent situational information gathering in the markov environment. *Ieee Systems Journal*, 14(2):1798–1809.
95. Zhu, S. J., Kim, W., Chang, G. L., and Rochon, S. (2014). Design and evaluation of operational strategies for deploying emergency response teams: Dispatching or patrolling. *Journal of Transportation Engineering*, 140(6):12.

5 Conclusion

In this study, we conducted a systematic literature review on patrolling-related problems. Although no year restriction was applied, the articles retrieved spanned from 2001 to 2024. We carefully screened the papers and categorized them to facilitate tracking their contributions, solution methodologies, and application domains. To present our findings in a more digestible format, we utilized tables and graphs. Ultimately, 122 journal articles were analyzed in this review.

Our analysis revealed that game theory is the most widely adopted methodology for modeling patrolling problems, largely due to its ability to capture strategic interactions between competing players (e.g., attacker-defender, smuggler-patroller). As a result, game-theoretical models were examined in greater depth, with additional categorization based on aspects such as information structure and rationality. Particular attention was given to studies that relaxed these two critical assumptions, offering valuable insights into their potential impact on future research. Additionally, we reviewed game-theoretical models based on the domains they addressed (e.g., theoretical, border security, transportation security), providing insights and statistics for each domain to better highlight their contributions. To the best of our knowledge, domain-based categorization has not been applied in prior reviews on this topic. The results also highlighted the significance of deterministic and stochastic optimization, both as primary and secondary modeling themes. Integrating multiple mathematical disciplines into the modeling framework provides richer insights and greater potential to address real-world scenarios. In particular, incorporating stochastic elements, while adding complexity and sometimes leading to intractable solutions, is expected to yield outcomes that are well worth the effort.

We deeply analyzed and categorized the directions that the authors of the papers in this study intended to explore (or suggested for professionals to investigate in the future).

Y. I. Tokel and J. Zhuang, *Optimizing Security Patrolling Strategies*, Synthesis Lectures on Operations Research and Applications, https://doi.org/10.1007/978-3-032-02617-0_5

Based on our classification, we identified ten recurring main themes mentioned in the future research directions sections. The most recurring and notable themes included ideas related to expanding their models, relaxing the assumptions made during model development, proposing algorithmic improvements when the capabilities of their solution algorithms or heuristics were found to be restrictive, and emphasizing the need to validate models through different case studies or human subject experiments. Alternative modeling approaches were also suggested when current models limited the quality of solutions. Some authors highlighted the importance of developing dynamic and online policies to incorporate real-time aspects into their proposed methodologies. Other recurring themes included the need for more data collection, relaxing the assumption of identical units in multi-unit studies, considering cooperation between units, and extending single-objective formulations to multi-objective ones. We shared our insights on the importance of these categories, referenced specific papers, and discussed their targeted suggestions along with how these can be implemented.

GPSR Compliance

The European Union's (EU) General Product Safety Regulation (GPSR) is a set of rules that requires consumer products to be safe and our obligations to ensure this.

If you have any concerns about our products, you can contact us on ProductSafety@springernature.com

In case Publisher is established outside the EU, the EU authorized representative is:

Springer Nature Customer Service Center GmbH
Europaplatz 3
69115 Heidelberg, Germany

Batch number: 09460790

Printed by Printforce, the Netherlands